"I strongly believe that the University of Toronto offered me the best education in mathematics and physics available anywhere in the world in the 1940's, when I was a BA and MA student there. ...

In light of my unexpected Nobel Prize in CHEMISTRY I find it amusing to remember that, in spite of Dean Beatty's strong intervention and my own feeble efforts, the Chair of Chemistry, Professor Kenrick, prohibited me and a few others from entering the Chemistry Building, because of our Austrian/German nationality, and thus excluded us from all college level Chemistry courses."

***

Walter Kohn, 1998 Nobel Laureate in Chemistry and discoverer of the Density Functional Theory (DFT), died in 2016 at the grand age of 93. This book is the first ever biography of Kohn, who led a remarkable life and scientific career, not least the fact that his DFT theory has emerged as the underlying computational method for molecular simulation used throughout the physical and life sciences. Taking us on a compelling journey, Sir David Clary traces Kohn's early life in Vienna and his dramatic escape from the Nazis on the Kindertransport to England in 1939, followed by Kohn's internment as an "enemy alien" and his transportation to Canada in 1940. His subsequent scientific career is discussed in detail, including his remarkable sabbatical in France when he discovered DFT, and his enduring efforts on peace initiatives and reduction of nuclear proliferation. An extraordinary story of a theoretical physicist winning the Nobel Prize in Chemistry, *Walter Kohn* is a sparkling chronicle of one of the great scientists of the 20th century who forever changed the way contemporary science is done.

Walter Kohn

From Kindertransport and
Internment to DFT and the
Nobel Prize

Other World Scientific Titles by the Author

Schrödinger in Oxford
ISBN: 978-981-12-5000-2
ISBN: 978-1-944660-28-4 (pbk)

The Lost Scientists of World War II
ISBN: 978-1-80061-475-8
ISBN: 978-1-80061-491-8 (pbk)

Walter Kohn

From Kindertransport and
Internment to DFT and the
Nobel Prize

David C Clary

University of Oxford, UK

World Scientific

NEW JERSEY · LONDON · SINGAPORE · BEIJING · SHANGHAI · HONG KONG · TAIPEI · CHENNAI · TOKYO

Published by

World Scientific Publishing Co. Pte. Ltd.

5 Toh Tuck Link, Singapore 596224

USA office: 27 Warren Street, Suite 401-402, Hackensack, NJ 07601

UK office: 57 Shelton Street, Covent Garden, London WC2H 9HE

Library of Congress Cataloging-in-Publication Data

Names: Clary, David C., author.
Title: Walter Kohn : from Kindertransport and internment to DFT and the Nobel Prize /
    David C. Clary, University of Oxford, UK.
Description: New Jersey : World Scientific, [2025] | Includes bibliographical references and index.
Identifiers: LCCN 2024017594 | ISBN 9789811291951 (hardcover) |
    ISBN 9789811292514 (paperback) | ISBN 9789811291968 (ebook for institutions) |
    ISBN 9789811291975 (ebook for individuals)
Subjects: LCSH: Kohn, Walter, 1923–2016. | Physicists--United States--Biography. |
    Jewish scientists--United States--Biography. | Nobel Prize winners--United States--Biography. |
    Density functionals.
Classification: LCC QC16.K636 C53 2025 | DDC 530.092 [B]--dc23/eng/20240509
LC record available at https://lccn.loc.gov/2024017594

British Library Cataloguing-in-Publication Data
A catalogue record for this book is available from the British Library.

For any available supplementary material, please visit
https://www.worldscientific.com/worldscibooks/10.1142/13806#t=suppl

Desk Editor: Shaun Tan Yi Jie

Typeset by Stallion Press
Email: enquiries@stallionpress.com

About the Author

Sir David Clary FRS is Emeritus Professor of Chemistry at the University of Oxford, UK. He was President of Magdalen College, Oxford from 2005–2020. He is an elected fellow of many academies including the Royal Society, the Royal Society of Chemistry, the Institute of Physics, the American Physical Society, the American Association for the Advancement of Science, the American Academy of Arts and Sciences, and the International Academy of Quantum Molecular Science. He was President of the Faraday Division of the Royal Society of Chemistry from 2006–2008. From 2009–2013, he was the first Chief Scientific Adviser to the UK Foreign and Commonwealth Office. In 2016, he was knighted in the Birthday Honours of Queen Elizabeth II for services to international science.

Sir David is a theoretical chemist recognised for his pioneering work on the quantum dynamics of chemical reactions. He has published over 350 papers in this field. He was Editor of *Chemical Physics Letters* from 2000–2020 and a Reviewing Editor of *Science* from 2003–2016. He has won many prizes for his research including the Royal Society of Chemistry Meldola, Marlow, Corday-Morgan, Tilden, Polanyi, Chemical Dynamics, Liversidge and Spiers awards, and the Medal of the International Academy of Quantum Molecular Science. His books *Schrödinger in Oxford* and *The Lost Scientists of World War II* were published by World Scientific in 2022 and 2024, respectively.

Preface and Personal Acknowledgements

Walter Kohn, the Nobel Laureate in Chemistry for 1998 and the pioneer behind the Density Functional Theory (DFT), passed away in 2016 at the age of 93. He has become one of the most influential scientists who worked in the second half of the 20th century. This biography describes Kohn's extraordinary life, a journey intertwined with remarkable scientific achievements. His creation, the DFT method, now stands as the cornerstone of computer simulations across all scientific disciplines involved with molecules or materials.

The biography presents the narrative of Kohn's early days in Vienna, his daring escape from Nazi persecution via the Kindertransport to England three weeks before the start of the war in 1939, his subsequent internment as an "enemy alien" and his transportation to Canada in 1940. The book describes Kohn's unusual education, received amongst the most difficult challenges, and his subsequent highly creative career as a theoretical physicist in the USA. Pivotal moments are highlighted such as his transformative sabbatical in France, where he unearthed the fundamentals of DFT and then derived the key equations to implement his theory. His early life in the turmoil of the late 1930s and early 1940s, together with the tragic fates of his parents, family members and teachers, gave him very deep convictions and the book describes his eloquent efforts for peace and nuclear disarmament. The story of Kohn's Nobel Prize in Chemistry, a rare honour for a theoretical physicist, is also discussed in detail.

Walter Kohn was a warm, humane and highly popular scientist. Accordingly, there have been a number of previous publications about his career including an excellent extended article by Andrew Zangwill, a book of *Personal Stories and Anecdotes Told by Friends and Collaborators* edited by Matthias Scheffler and Peter Weinberger, and several reviews of his work, biographical memoirs and obituaries. Following on from his first academic appointment to the Carnegie Institute of Technology in 1950 there are extensive archives of Walter Kohn's papers, which are kept at the University of California, Santa Barbara, and this collection has been invaluable for the biography. After the award of his Nobel Prize in 1998, he gave numerous interviews about his very challenging life in the dangerous times of the late 1930s and early 1940s, and this has also provided essential information for the book.

I would like to give special thanks to many individuals, archives and institutions who provided permissions and assistance for the material included in this biography, including photographs, letters and quotes. Details on permissions are given in the figure captions and in the additional summary at the end of the book. In particular, I would like to thank Ingrid Kohn Paymar (daughter of Walter Kohn), Josef Eisinger (Kohn's close friend who celebrated his 100th birthday when this book was being written), Hilary Pople (daughter of John Pople), Jeanne Parr Lemkau (daughter of Robert Parr), Peter Neuhaus (son of Kohn's schoolfriend Herbert Neuhaus), and Geert-Jan Kroes for their very helpful assistance.

A small number of letters and quotes have been translated by the author and are denoted in the text by (t).

Front cover credit: Adapted from an electron density map for the hydrogen molecule calculated with Density Functional Theory (courtesy of John Wonmo Seong), together with photographs of Walter Kohn (courtesy of the family of Walter Kohn).

Contents

Chapter One

Vienna

On 13 October 1998 at 5 in the morning Mara, the wife of Walter Kohn, took a telephone call at their home in Santa Barbara, USA. A pleasant voice said this was from the Royal Swedish Academy of Sciences. Mara's jaw dropped. She could not say anything. She gave the phone to her husband. He was told that he had been awarded the Nobel Prize for Chemistry together with John Pople.[1]

Then the phone in Sweden was passed to Björn Roos, a theoretical chemist from Lund who was a member of the Nobel Prize Committee for Chemistry. Roos said he needed to emphasise that the Nobel Prize was in Chemistry, not Physics. Walter replied, "Well, that, to me, makes it a lot more fun. I mean it's hard enough for a physicist to get the Nobel Prize in Physics, but to get it in Chemistry is even a little more unusual."[1]

Before this date, Walter Kohn had only published one research paper in a scientific journal concerned directly with chemistry. However, his own fundamental contribution had changed the way chemists and other scientists perform computations on molecules and materials. He had an extraordinary personal story that led to the Nobel Prize. This story is told in this book.

Walter Kohn was born on 9 March 1923 in Vienna, Austria to a relatively prosperous Jewish family.[2] His father Salomon Kohn came from Veselí, Moravia, a small town in what is now the Czech Republic and quite close to the border with Austria.[3] His birthdate was 31 July 1873. Salomon had a successful business in publishing artistic postcards. He was first married to Marie Kohn (née Katay), who died in 1915. Walter's mother was Gittel (née Rapaport, sometimes spelt Rappaport), who was Salomon's second wife and was born on 18 May 1892 in Brody in the Galician region of Poland, which is now in Western Ukraine.[4] The Rapaports were a large family and gave the nickname "Rappa" to Walter that was used by some of his school friends even after he moved to North America.

Both Walter's parents had come to Vienna in the time of the thriving Austro-Hungarian Empire during the late 1800s.[5] His father spoke some German and Slavic languages but not English. Walter described his mother Gittel as "a small, good girl" who "was highly educated, including Latin and Greek." She spoke English and French in addition to German, and "was comfortable as a Jewish woman and also in living subsequently in a German-speaking country."[6] Salomon was 19 years older than Gittel and Walter commented that his father always seemed much older than his mother.

Gittel's parents, Pincus and Elizabeth Rapaport, were Orthodox Jews who moved with her to Vienna. They had two daughters (Gittel and Mala) and three sons (Dario, Joseph and Samuel) (see Figure 1). Pincus and Elizabeth got to know their grandson Walter quite well and he remained proud of his deep Jewish heritage.[6]

Walter had an elder sister Minna, sometimes called Minnie, who was born on 29 August 1919.[7] She also had an interesting life story that will be discussed in later chapters.

Before the First World War, Vienna was a city with a very creative society and a thriving cultural scene which attracted the entrepreneurial

Fig. 1. The Rapaport family. Left to right: Samuel (brother), Gittel by the piano, Pincus (father), Joseph (brother), Dario at the back (brother), Elizabeth (mother), Mala (sister). (Courtesy of the family of Walter Kohn.)

Fig. 2. Theobaldgasse 7, Vienna. Home of the Kohn family. (Courtesy of Hedwig Abraham.)

Salomon Kohn.[8] The family lived in Theobaldgasse 7, a fine building in the Vienna 6th District off the Gumpendorferstrasse and quite close to the Viennese State Opera (see Figure 2). Theobaldgasse had been the home of Haydn, Mozart and Beethoven and this deep musical heritage was important to Walter Kohn throughout his life.

In Vienna, Salomon Kohn founded a postcard publishing company in 1898, Brüder Kohn, together with his two brothers Adolf and Alfred.[9] They printed iconic art postcards which had the distinctive signature from Wien District 1: "BKW1". Picture postcards with images on one side were first created in the 1870s in France and the Kohn brothers were able to exploit this popular new colour medium in Vienna. A feature of many of the cards were paintings and motifs of unusual places and buildings in Vienna, commissioned from talented artists. They included pictures, and sometimes satirical caricatures, of notable people such as actors, composers and politicians. The business flourished and branches were opened in Teinfaltstrasse, Kärntnerstrasse and Mariahilferstrasse in Vienna. A branch was even opened in Friedrichstrasse in Berlin.[9]

Fig. 3.   A satirical postcard from Brüder Kohn. "In the valley of tears" by Tony Pfoser.

The Viennese State Association for Tourism was keen to promote the postcards and featured a highly acclaimed exhibition of the Brüder Kohn work in 1912.[10] The postcards have remained collectors' pieces to the present day and reflect the artistic and liberal atmosphere of Vienna in the period before the First World War (see Figure 3). With the success of the company, Salomon Kohn's apartment on Theobaldgasse became a meeting place for many of the most notable people then present in Vienna including Gustav Mahler, Max Reinhardt, Arthur Schnitzler and Enrico Caruso. Politicians such as the social democrat Karl Seitz, who acted briefly as President of Austria after World War I and became Mayor of Vienna, were also frequent visitors.[3]

Fig. 4.   The Kohn family on vacation in Heringsdorf on the Baltic sea, ~1925. Walter (rightmost) is directly in front of father Salomon, and Minna is in front of mother Gittel (leftmost). (Courtesy of the family of Walter Kohn.)

Walter Kohn never forgot this liberal heritage that once thrived in his family home in Vienna (see Figures 4 and 5). However, there were challenges after World War I and the fall of the Austro-Hungarian Empire. In his own words:

The [postcard] business had flourished in the first two decades of the century but then, in part due to the death of [my father's] brother Adolf in World War I, to the dismantlement of the Austrian monarchy and to a worldwide economic depression, it gradually fell on hard times in the 1920s and 1930s. My father struggled from crisis to crisis to keep the business going and to support the family. Left over from the prosperous times was a wonderful summer property in Heringsdorf at the Baltic Sea, not far from Berlin, where my mother, sister and I spent our summer vacations until Hitler came to power in Germany in

1933. My father came for occasional visits (the firm had a branch in Berlin). My mother was a highly educated woman with a good knowledge of German, Latin, Polish and French and some acquaintance with Greek, Hebrew and English. I believe that she had completed an academically oriented High School in Galicia. Through her parents we maintained contact with traditional Judaism. At the same time my parents, especially my father, also were a part of the secular artistic and intellectual life of Vienna.[11]

Fig. 5.   A happy Kohn family, ~1932. Left to right: Walter, Minna, Gittel, Salomon. (Courtesy of the family of Walter Kohn.)

Fig. 6.   Akademisches Gymnasium, Vienna. (Courtesy of C.Stadler/Bwag; CC-BY-SA-4.0.)

Salomon Kohn was a pacifist and did not serve in World War I. However, his brother Adolf volunteered for the Austro-Hungarian army and was killed in 1918. Despite the financial challenges to the Brüder Kohn business after the war, Walter was enrolled at the Akademisches Gymnasium in 1933 at the age of ten (see Figure 6).[6] This was the oldest secondary school in Vienna. Founded in 1553 by Jesuits, the school had a reputation for a humanist education.[12] Originally, the school only taught religious subjects but after the Jesuits were dissolved by Pope Clement XIV in 1773 the range of subjects was enlarged significantly. By the late 1800s, the school, now located at the Beethovenplatz, had become the favoured institution for the children of the liberal and cultural elite of Vienna, which now had a high Jewish representation. By the early 1900s, Catholic and Jewish children were almost equally represented at the school. Many famous Viennese had been educated there including Franz Schubert. The teaching of the Classics of Latin and Greek had become particularly strong at the Gymnasium by the 1930s with less emphasis on science and mathematics. Nevertheless, several famous Austrian physicists had been pupils there including Ludwig

Fig. 7. Walter Kohn with his mother Gittel, ~1936. (Courtesy of the family of Walter Kohn.)

Boltzmann, Paul Ehrenfest, Richard von Mises and Erwin Schrödinger. As we shall see, the link to Schrödinger was particularly important to Walter Kohn in more ways than one.

Before World War II, the teaching in the Akademisches Gymnasium was for boys only but the most promising female students in Vienna were allowed to take their examinations there. This included the pioneering physicist Lise Meitner who described the formidable challenge of passing the Matura examination in 1901 alongside Henriette Boltzmann, the daughter of the great physicist, to enable her to attend the University of Vienna.[13] The Jewish and Catholic pupils were segregated in classes but Walter commented that one of his best friends was a Catholic with a baptised Jewish father and they subsequently retained contact throughout their lives.[6] Nevertheless, Walter and his sister Minna were brought up under the Jewish tradition which they picked up particularly from his mother and her parents, and he had his Bar Mitzvah (see Figure 7).

Walter Kohn wrote positively about his education at the Akademisches Gymnasium (see also Figure 8):

Fig. 8. Walter Kohn at the age of 14 in the snow of Vienna, 1937. (Courtesy of the family of Walter Kohn.)

There for almost five years I received an excellent education, strongly oriented toward Latin and Greek… my favorite subject had been Latin, whose architecture and succinctness I loved. By contrast I had no interest in, nor apparent talent for, mathematics which was routinely taught and gave me the only [grade] C in high school. During this time it was my tacit understanding that I would eventually be asked to take over the family business, a prospect which I faced with resignation and without the least enthusiasm.[11]

The year that Walter started at the Akademisches Gymnasium proved to be one of the most ominous in the 20th century. Following economic and political upheavals, Adolf Hitler was appointed Chancellor of Germany on 30 January 1933 by President Paul von Hindenburg. Almost at once, anti-Semitic laws were introduced in Germany by Hitler and his National

Socialist Party. The Law for the Restoration of the Professional Civil Service was introduced that prevented "non-Aryans" from holding positions in the Civil Service.[14] This included many academics in the German universities. Someone whose parent or grandparent was of Jewish blood or religion was categorised as a non-Aryan. These laws were soon extended to professionals such as lawyers and doctors, and then the Nuremberg Race Laws were introduced in 1935 which removed the political rights and full German citizenship of Jewish people.

In Austria, these developments in Germany were observed with glee by members of the *Nationalsozialistische Deutsche Arbeiterpartei* which had long been campaigning for the unification of Austria with Germany. Hitler was born in Braunau am Inn in Austria and had written in his book *Mein Kampf* that there must be an Anschluss — a union created between Austria and Germany. Over many years there had been a considerable resentment built up in Austria against the increasing number of Jewish people who had migrated to Vienna and other Austrian cities from the Eastern regions of the Austro-Hungarian Empire in the period before the First World War.[8] The Kohn family is an example of such a migration. The humanist emphasis of the German-speaking world, as exemplified by Goethe, Kant and Mozart, was taken up with enthusiasm by the Eastern Jews whose previous cultural background had been highly limited. In Vienna during this period, over half the doctors and lawyers were Jewish as were many of those making artistic contributions.[8] The Jewish population in Vienna increased from 4,000 in 1847 to over 175,000 in 1910.[8] Some of these Viennese Jewry discarded their previous traditional and highly religious lifestyles and thrived in the creative and Germanic culture of Vienna. However, these major changes were observed with envy by some of the non-Jewish inhabitants of the city. Hitler himself was a homeless artist in Vienna in 1908 and his resentment towards Jewish people can be traced to this period.[15]

Following World War I, the Christian Socialist Party played a leading role in Austria and did not support an Anschluss. Its leading members included Karl Seitz, the Mayor of Vienna and close associate of Salomon Kohn. However, after Hitler was made Chancellor of Germany in 1933 the Chancellor of Austria, Engelbert Dollfuss, ruled by decree and many political parties were banned. Following the assassination of Dollfuss by the Nazis in 1934, Kurt Schuschnigg was made Chancellor. The pressure on

Austria continued to grow and on 12 February 1938, following a personal meeting with Hitler at Berchtesgaden, Schuschnigg agreed to appoint the Nazi sympathiser Arthur Seyss-Inquart as Minister of Public Security and in charge of the police. However, after Schuschnigg called a referendum proposing to keep Austria separate from Germany the furious Hitler ordered his troops to invade Austria on 12 March 1938 and the Anschluss was essentially complete. Hitler himself then went to Vienna and announced from the balcony of the Hofburg Palace to over 200,000 ecstatic supporters in the Heldenplatz that the country of his birth had been incorporated into the Third Reich. There followed the quick arrest by the Schutzstaffel (SS) of prominent Jewish people, leading politicians and other Nazi opposers under the direction of Heinrich Himmler and Reinhard Heydrich.[16]

School pupils at the Akademisches Gymnasium at the same time as Walter stated that they did not come up against anti-Semitism at the school before 1938, although some teachers, including the teacher of geometry, were known to be active Nazis. Walter commented that he did not himself see any anti-Semitism before the Anschluss and, for him, Vienna was a liberal and progressive city.[6] Before 1938, life was happy for the Kohn family in Vienna (see Figure 9).

Fig. 9.   Walter Kohn and his sister Minna, 1937. (Courtesy of the family of Walter Kohn.)

However, things changed very quickly in March 1938 and some students were attacked by Nazi supporters on their way to school. Walter commented:

> After the Anschluss I was walking home with a cast on my leg after a skiing accident. I was surrounded by Nazi hooligans who were protected by policemen with swastikas on their uniforms. I managed to get home after losing a shoe and my parents were distraught. They realised at once they had to get their children out of Austria.[6]

Walter Kohn was just past his 15th birthday at this time and observed closely the tumultuous developments in Austria which he would never forget. He stated: "Hitler's speeches were just screams. Vienna before Hitler was the centre of European civilisation with significant contributions from Jewish people. My family could not imagine how Vienna could be taken over by this screaming man."[6]

Following their recent experiences in Germany, the Nazis were able to impose new laws in Austria at a startling speed.[16] The Director of the Akademisches Gymnasium Dr Ludwig Marcus was Jewish and was removed from his post.[12] He was replaced by the National Socialist Dr Hans Schmidt who, within weeks of the Anschluss, forced the expulsion of nearly half of the school pupils, including Walter and three teachers who were Jewish. Walter said that the Gymnasium was closed for a few weeks after the Anschluss and he returned briefly to the school. Then the Jewish students were all notified that they had been expelled.[6]

Walter's education had frequent major interruptions in Austria, England and Canada but each change introduced unexpected aspects which were to prove surprisingly positive for his academic development. Some of the students from his class at the Akademisches Gymnasium never went to school again but, overall, he had achieved good grades which enabled him to be enrolled in August 1938 at the Zwi Perez Chajes Gymnasium (Figure 10). This was a private school for Jewish students that was named after a Jewish rabbi.[17] This school, for both boys and girls, was established

Fig. 10.   Chajes Gymnasium in Vienna. (From Das Chajesrealgymnasium in Wien 1919–1938 by B. Shimron. Courtesy of Leo Baeck Institute, New York.)

in 1919 in the Vienna 2nd District at Castellezgasse 35, near the Augarten. After the Anschluss it was the only high school in Vienna that was able to accept Jewish students. The school was highly selective and, with very bright Jewish students being expelled from all over Vienna, the competition for places in the summer of 1938 was strong.

The Chajes Gymnasium had a reputation for a very good organisation and spirit throughout the school. The school was proud that it always started at 8 am, timetables were adhered to, and work was always handed back marked and on time. The head of the School Viktor Kellner himself taught a wide range of subjects from philosophy to German literature and students described this as an "overwhelming experience" and he was "able to turn a lesson into an intellectual exercise".[17]

The influence of the Nazis on the Chajes Gymnasium was immediate after the Anschluss. At the graduation ceremony of 1938 swastika flags had to fly on the school buildings and Sturmabteilung (SA) men stood guard. The school head Viktor Kellner spoke bravely: "I don't know what future you have ahead of you. But I can tell you for sure, people will say Schma Israel longer than Heil Hitler!"[18] With the subsequent forced departure of many

Fig. 11.   Emil Nohel, ~1938. (Courtesy of the Yad Vashem Collections.)

Jewish students from Austria, the number of students in the school quickly reduced. It was strong in teaching music and German but also science. Walter was a pupil there for just one year but this initiated his interest in physics. Kellner left for Palestine in October 1938 and was replaced by Dr Emil Nohel (see Figure 11).

Emil Nohel had a humble background.[19] He was born in 1886 to a Jewish family in the village of Mcely in Bohemia, which is now in the Czech Republic. He was educated in the German University of Prague. This university was establishing an excellent reputation in mathematics through individuals including Ludwig Berwald, Paul Funk, Walter Fröhlich, Karl Löwner, Erwin Finlay Freundlich and Georg Pick.[20] Nohel's tutor Professor Anton Lampa told him in 1904 not to take physics as a main subject "since all the original work had already been done, the laws had been established, and important new developments were not to be expected."[19] However, on Lampa's recommendation, Einstein took Nohel as a paid research assistant ("Wissenschaftliche Hilfskräfte") when he moved to Prague in 1911. By this time, Einstein had already made his reputation for his revolutionary research on special relativity, Brownian motion and the photoelectric effect. There is no record of the scientific interaction between Nohel and Einstein but Nohel's son wrote: "The many hours Einstein and my father spent together in Einstein's study, his world view and character left a lasting impression."[21]

Being a research assistant to Einstein was highly prestigious for Nohel not just because of the fame of his supervisor but through others who took up his post. Nohel's predecessor was Ludwig Hopf, who contributed to the development of the Special Theory of Relativity and published with Einstein on the statistical theory of radiation. Nohel's successor was Otto Stern who discovered electron spin and was awarded the Nobel Prize for Physics of 1943 for the first measurement of the magnetic moment of the proton (and who would become a colleague of Walter in the Department of Physics at the Carnegie Institute of Technology in the 1950s). This fundamental experimental work led to the hugely important techniques of Nuclear Magnetic Resonance and Magnetic Resonance Imaging.

Einstein returned to Zurich in 1912 and Nohel then completed his doctorate in mathematics under the supervision of Georg Pick. With Anton Lampa, Pick had also been a member of the committee which appointed Einstein and he contributed to Einstein's development of General Relativity. In 1914, Nohel wrote a paper "On the natural geometry of plane transformation groups" which was published in the academy journal, the *Sitzung-Berichte der Akademie der Wissenschaften in Wien*.[22]

After this, Nohel taught mathematics at the Vienna Handelsakademie in buildings which now form the Vienna Business School. Following the Anschluss he was expelled from the Handelsakademie and joined the Chajes Gymnasium first as a teacher and then Director.[17] There he taught Walter elementary physics and mathematics, which Walter very much appreciated and frequently mentioned in interviews. Walter only learnt some ten years later that Nohel had been an assistant to Einstein.[23,24]

In 1938–9, keeping the Chajes Gymnasium going was a major challenge for Nohel. Walter reported:

> We started the regular school day, and we were told, well, unfortunately, our teacher of Latin had disappeared since the previous day. Of course, he was arrested and taken away (by the Gestapo). Well, there wasn't a different or substitute teacher available, and so that same Professor Nohel whose

field was physics, said: "Well, we don't have anybody to teach you Latin, so I'm going to do it. I haven't had any Latin for the last twenty years, but with a dictionary, I think I can do it." So, if I remember correctly, we read Virgil together, and in some way, you know, under such stressful conditions, in a way, you appreciate more what you learn, because you say to yourself: "Well, the whole thing may stop tomorrow, may be the end of it."[1]

Herbert Neuhaus was a student at the Chajes Gymnasium in Walter's class. He wrote on the unique atmosphere in the school:

There was a strong bond of comradeship, even friendship, between students and professors, and there was only one common goal: to learn as much and as quickly as possible. The official curriculum was thrown out of the window and we pushed ahead up to and beyond university level. I remember that some of us went after school and after the school year had ended to the home of our mathematics teacher, Prof. Viktor Sabbath on Währinger Strasse to continue in higher mathematics — until he was deported to a death camp.[25]

Walter also wrote very favourably about the teaching of Viktor Sabbath. He recalled that Sabbath introduced him to the book *Matter and Light* by the Nobel Physicist Louis de Broglie.[1] It is extraordinary that so many of the students in Walter's class, all born in 1923, went on to outstanding scientific careers in other countries after harrowing experiences or escapes in the Second World War. Gertrude Ehrlich, Rudi Ehrlich (no relation to Gertrude) and Karl Greger all became distinguished professors of mathematics. After being one of the small number of people to survive the Theresienstadt concentration camp, Herbert Neuhaus eventually took up the

Fig. 12. Students at the Chajes Gymnasium in 1938. Back row from the right: Herbert Neuhaus, Walter Kohn, Paul Sondhoff. Second row far right: Rudi Ehrlich (later Permutti). First row far right: Gertrude Ehrlich, and next to her, Ilse Arnold-Levai. (Courtesy of Peter Neuhaus.)

post of Medical Director at the Illinois Public Health Hospital in Chicago and was able to meet up with Walter and other friends for a class reunion in 2000, over 62 years after they were at the Chajes Gymnasium together.[25]

The photograph of Figure 12 shows the schoolmates of 1938–9 with Walter in the back row with his arms around Herbert Neuhaus and Paul Sondhoff. After being hidden in Vienna during the war, Sondhoff became an aeronautical engineer working for the US government (see also Chapter Five).

After the Anschluss in 1938, managing the Brüder Kohn postcard business became a huge challenge for Walter's father Salomon. He was frequently hassled by Nazi members and eventually one came to him and said he was taking over ownership of the business, while Salomon had to stay working without pay. Walter said:

My father was forced to give up the business to an individual who had no experience of the business and had been a member of the Nazi SA even before the Anschluss. My father had no say in the very small sum which he was paid for the business. The new owner was not able to run the business and ordered my father to do it. My father felt, in the short term, he had no other option… My father got a letter from the foreign office informing him that because he was essential for maintaining the business he couldn't leave Austria. He would not receive an exit visa. He would be hired as the manager, but it was pay that didn't exist.[23]

The November Pogroms (often called Kristallnacht) of 9–10 November 1938 were highly traumatic for Jewish people throughout Nazi-controlled Germany, Austria and the Sudetenland.[26] Following the shooting of a German diplomat Ernst vom Rath in Paris by a 17-year-old Polish Jew Herschel Grynszpan, the SS and SA paramilitary organisations went on a rampage of destruction and arson of Jewish businesses, homes and synagogues. Numerous Jews were attacked and some were murdered. Walter personally recalled these dreadful assaults:

In the original Kristallnacht, a friend and I just stepped out of this Jewish school and this was reason enough for us to be taken by a seemingly very friendly Austrian policeman to a police station and to be held there for many hours, terrified. I came home and found our apartment absolutely vandalized by a group of hooligans, including the person who had taken over my father's business ... My father had a collection of fine Viennese glass in a case. They just reached in and tossed each piece into a mirror in the room. Kristallnacht was exactly the right name for this.[6]

Like many other Jewish men in Vienna, his father had been arrested on Kristallnacht and came home to find the destruction in his home. Walter had also witnessed the humiliation of Jewish people being forced to scrub the streets on their hands and knees while being watched by laughing "Nazi hooligans", in Walter's own words.[6] Some Jewish people had already started to emigrate from Vienna before the Kristallnacht. It was the warning sign for Walter's parents to find a way urgently for their children to leave Austria.

Chapter Two

Kindertransport

The Kristallnacht was a wake-up call not just for the regions controlled by the Nazis but for the whole of Europe. It was the first time there had been systematic and organised attacks on Jewish people and property on a very wide scale. There were over 1,000 synagogues set on fire and the fire brigades were only called out to protect neighbouring "Aryan" properties. It was followed by the arrest of over 30,000 Jewish men and political dissidents who were sent to concentration camps such as Dachau and Buchenwald.[26]

With the alarm bells ringing, many of the Jewish community in Vienna at once looked to see if it would be possible to leave the country. This was the case for the younger members of the Kohn family. The Israelitische Kultusgemeinde Wien (IKG Wien) was the organisation set up by order of the Nazi authorities in May 1938 through its Central Office of Jewish Emigration. The IKG was run from a property in Vienna previously owned by Louis Nathaniel de Rothschild. Before March 1938, there were 6,900 Jewish children aged up to six, 7,600 between six and 14, and 4,500 aged between 14 and 16, living in Vienna.[27]

It was the idea of Adolf Eichmann that forcing Jewish groups to assist with making arrangements for emigration would be a subtle method having more chances to succeed. Eichmann had set up a similar office to IKG Wien in Berlin in 1935 with the aim of expelling Jewish people as quickly and efficiently as possible by forcing the leaders of the Jewish communities to cooperate. After the start of the war, these organisations were made to be involved in more ruthless activities including the confiscation of Jewish assets and, eventually, the establishment of lists of Jewish people who would be transported to concentration camps.[28]

Within a week of Kristallnacht, on 16 November 1938, the 15-year-old Walter Kohn made his first written application to the IKG. He stated on his application form that he was hoping to emigrate to Holland.[29] However, this application did not at once proceed as he needed to get permission from the Dutch authorities to enter their country. At that time there was no special scheme set up by other countries for the emigration of children from Austria or Germany.

In the UK several individuals and groups at once realised that the situation in the Nazi-controlled regions had become critical. Just six weeks before the Kristallnacht, Hitler and Mussolini had their fateful meeting with the UK Prime Minister Neville Chamberlain in Munich that enabled Hitler to send his troops into the Sudetenland of Czechoslovakia. The Munich agreement gave the impression to the Nazis that the great powers were not concerned with their horrendous actions. These started on a major scale with Kristallnacht and then moved on in due course to the death and destruction of the Second World War and the Holocaust. Chamberlain had returned triumphantly to London announcing "Peace in our Time" but many in his country saw things differently.[30]

Within five days of Kristallnacht a prominent British Jewish delegation including Viscount Samuel, who had been the leader of the Liberal Party between 1931 and 1935, and also Lionel Rothschild and Chaim Weizmann, had a meeting with Prime Minister Chamberlain in which they proposed admitting 10,000 Jewish children under 17 into the UK from the Nazi-controlled countries of Europe.[31] It seems that Chamberlain himself did not reply with enthusiasm to the proposal but his Cabinet, which met next day, had a more positive view. A full-scale debate then took place in the House of Parliament on the matter on 21 November 1938. The Home Secretary Samuel Hoare came from a family with a long Quaker tradition. He announced that the Home Office would allow entry of child refugees providing their maintenance was guaranteed.[32] Lord Halifax the Foreign Secretary, who had been aligned with the appeasement movement, also backed the scheme hoping that this would bring the USA into opposition of the Nazis. The Government then agreed on 23 November that an unspecified number of children under the age of 18 would be allowed to enter the UK for their education and for a period of two years. Preference would be given to boys who would work in the agricultural sector.[27] This was to help Walter Kohn when he left Austria.

The "Kindertransport" (children's transport) was a major change in UK government policy, initiated at untypical speed.[33] The total number of Jewish people who were able to emigrate to the UK was estimated at about 10,000 for the five years before March 1938, and this figure was then increased fivefold for the subsequent period up to the start of the war in September 1939.[34] An imposed financial condition resulted in £50 having to be paid in advance for each child admitted.[27] This was a non-trivial sum, equivalent to £4,000 in 2023. Accordingly many of the children on the Austrian Kindertransport came from the better-off families of Vienna. The initial aim of the £50 was to provide funds to send the child back to their home country when the crises were over. It was assumed mistakenly that the Nazis would not last for long.[33]

Initially the Jewish community in the UK were confident that the funds could be raised to support the children. The Rothschild family announced an appeal for £1m to support "1,000 homeless orphans".[33] However this confidence was misplaced as the number of children transported to the UK increased. In the summer of 1939 the government took over the scheme when it was realised that the many thousands of children would not be able to return to their home countries. The UK body initially responsible for organising the Kindertransport was the Movement for the Care of Children from Germany, which quickly changed its name to the Refugee Children's Movement when children from Austria and Czechoslovakia started applying.[27] With their local links across the UK, arrangements were made to host the refugee children with families, in hostels and even in holiday camps. Most made the long distance from Europe on trains from Berlin, Vienna and Prague to Holland, and then by boat via the Hook of Holland to Harwich, and onwards to Liverpool Street Station in London for dispersion throughout the country.[35]

The scheme had the enthusiastic support of the British public. Former Prime Minister Stanley Baldwin raised the considerable sum of £500,000 in an appeal for Jewish refugees. Advertisements were placed in the leading newspapers with the slogan "Before it is too late, get them out!" and messages including "Our Christian Charity is Challenged" (see Figure 13). The first Kindertransport arrived within three weeks of Kristallnacht on 2 December 1938 with 200 children from Berlin.[33] There were not many praiseworthy actions from the UK government in the years running up to the Second World War but support for the Kindertransport was certainly one of them. Walter Kohn himself frequently remarked on his positive views of the scheme. British enthusiasm for the Kindertransport also led to some other immigration arrangements being introduced, including one for young

Shall they live?

BEFORE IT IS TOO LATE

get them out!

TEAR OUT THIS FORM NOW

LORD BALDWIN FUND FOR REFUGEES

Fig. 13. Appeal to support the Kindertransport from Lord Baldwin, 13 January 1939, Daily Mirror. (Courtesy of Reach Publishing Services.)

women who could undertake domestic service.[27] This was an entry point for Walter's elder sister Minna, who was too old to be part of the Kindertransport.

The Kindertransport had the tacit approval of the Nazi authorities who, at that time, were keen to remove as many Jewish people as possible from the countries under their control. However, the negotiations were complicated. A hugely courageous Dutch woman, Geertruida Wijsmuller-Meijer, travelled to Vienna to conduct negotiations directly with Eichmann for the Kindertransport to take place.[27,36] He imposed highly bureaucratic conditions for children to leave Austria, with complicated forms needing detailed information and the appropriate stamps from different departments that required queuing for several days for their completion. Furthermore the Nazis forbade overtly emotional displays and often insisted that the trains with the children should leave from non-central stations and after dark with the minimum of fuss from deeply worried parents and relations. Many parents thought that their children would be returned quite soon and few anticipated that this would be the last time they ever saw them. Very sadly, this was the case for Salomon and Gittel Kohn.

Lola Hahn-Warburg at the Refugee Children's Movement was the lead person taking responsibility at the London end for the children coming from Vienna. She insisted on children in good physical shape and strong mental health with a medical examination undertaken just before the transport left Vienna.[27] In addition, high educational attainment was necessary as evidenced through an appropriate school report. This tended to favour children from the better-off families in Vienna. These stipulations were in contrast to the children chosen by the IKG to have priority where those whose parents had been arrested by the Nazi authorities were put to the top of the list. In 1939, with children from Prague and some parts of Poland also needing assistance, tremendous pressure was placed on the Kindertransport scheme. With Hitler and the Nazis entering Prague in March 1939, that city had priority for the coming months and Walter was then still in Vienna hoping for a train. In February 1939, Lola Hahn-Warburg stipulated that children would only be accepted on the Kindertransport from Vienna if they were born after 1 March 1923. Walter's date of birth was 9 March 1923 so if he had been born just ten days earlier he would not have been allowed on the Kindertransport train to England.[27]

The experiences in the UK of the children who travelled via the Kindertransport were mixed.[34,35] A fair number, like Walter, were fortunate to be looked after by loving families. However, there is evidence that some were forced to undertake long hours of domestic work for the families they joined and there was even a very small number of examples of child abuse. At that time, private funds were needed in the UK to pay for a child's education over the age of 15 and, consequently, with the considerable pressure on funds, the Kindertransport was the end of the education for some of the children who, just months before, had been anticipating further studies back home.[37]

Werner Rosenstock had worked in Berlin for the Reichsvertretung der Juden in Deutschland (Representation of Jewry in German) from November 1938 to the start of the war and had been involved in the details of emigration. On 9 November 1958, the 20th anniversary of Kristallnacht, he wrote:

> The frontiers of most countries were barred. The United States and Palestine were restricted in their immigration policies by the quota and certificate systems. The only country which really reached out a helping hand, and which thus lived up to the emergency, was Great Britain. Of the 100,000–150,000 Jews who left Germany between the pogroms and the outbreak of war, about 40,000 found refuge in this island, and in addition a further 40,000 from Austria and Czechoslovakia. One must have experienced what it meant in those days of anxiety if a letter from a guarantor or from a British immigration authority arrived in a Jewish household.[34]

It was against this challenging background that Salomon and Gittel Kohn had decided to arrange for their children to go to the UK. Following the horrors of the Kristallnacht, Walter's father had written to a business associate Charles Hauff who lived in Copthorne in West Sussex in the UK.[6] Charles Hauff, who was not Jewish, had worked for a family company of the same name started by his father Carl Hauff (c1835–1905) who was born in Germany.[38] The company dealt with picture frames, prints, photographs, and fine art. They also published comical colour postcards which would probably have been noticed by Brüder Kohn. The Hauff company was known for being the first to introduce Florentine frames to England. From 1927–1938 the company was based at 42 Museum St. London. Charles Hauff had voluntarily liquidated the company in the summer of 1938 and was then living with his wife Eva at The Border Cottage in Copthorne, Sussex which was a small town 34 miles south of London on the Weald. Living in the house were also three elderly sisters of Charles Hauff and one brother.[39]

Salomon Kohn had not personally met Charles Hauff. He wrote cautiously to explain the "changed circumstances" in Austria.[23] He stated he would be "immensely grateful" if Mr Hauff and his family could find space for Minna and Walter for a short period. Hauff replied positively within one week. Minna was the first to leave. Her first application had been made to the IKG on 13 December 1938.[40] She wrote that she would prefer to go to Norway or Argentina. She stated she was studying to be a fashion designer in Vienna and spoke English and French. However, at this time she stayed in Vienna. Minna applied again on 6 March 1939. At an interview she stated that her parents were having to sell their furniture to survive.[40] She also stated she had been given a permit to be a domestic servant in England. This time her application was successful. Walter wrote to Minna on 21 April 1939:

> Dear Minnie! You can hardly imagine how happy I am that everything was successful for you. I wouldn't have believed that you could do everything so well in such a short time. So it now seems that I can finally get away, even though I still have doubts.[41(t)]

The Hauff household in Copthorne in 1939 already included six adults, some of whom had come to the house to avoid the expected Blitz of London.[39] Consequently, Eva Hauff would have welcomed help with domestic chores. On a subsequent UK "Female Enemy Exception of Internment Report" of 3 November 1939, which granted her exemption from all restrictions, a tribunal stated that Minna Kohn's nationality was Austrian, German by annexation. It indicated also that her normal occupation was a fashion designer while her present occupation was domestic service for her employer Mrs Eva Hauff.[42]

Walter was close to his cousins Georg and Lilly Kohn who lived nearby in Vienna and whose father was Alfred, one of the three brothers of Brüder Kohn. In 1938, the cousins were aged nine and 12, respectively. Walter's aunt Hilda Kohn had an offer of domestic service in England in 1938 but at that time she could not see a way to bring Georg and Lilly with her — she was

not prepared to send them on the Kindertransport on their own.[27] This was to be a fateful decision.

Walter Kohn was four years younger than his sister Minna and this made him just eligible for the Kindertransport. He was able to leave Vienna, aged 16, on 8 August 1939 with 48 other children.[43] Walter stated that his departure from the Western Station in Vienna was very painful for his parents.[6] However, he was older than most of the other children on his Kindertransport train and he already knew from letters from his sister that he would be looked after in England by a caring family. Many of the other children did not even know where they would be living in England.[27]

The train from Vienna went along the Rhine to Cologne. There Walter thoughtfully bought some Eau de Cologne as a present for his new family in England. At the Hook of Holland Walter recalls being given some "wonderful orange juice by a very kind lady".[6] After taking a boat he arrived at the port of Harwich, England on 10 August 1939, two days after leaving Vienna. He then travelled by train to Liverpool Street Station in London and on to Victoria Station where he was met by his sister Minna. She had quickly become fluent in English and took her brother to a café to have coffee and apple pie.[6] It was just one hour on a further train to Three Bridges, near the Hauff's house in Copthorne.

Walter said he at once was made to feel he was part of the Hauff family. Space was tight at the Border Cottage with several family members living there, but the Hauffs found an attic for Walter and Minna over their garage. It was primitive accommodation with gas lighting, no heating and no electric light but they were very grateful.[6] Mr Hauff was still busy with matters associated with his business and his wife Eva took full responsibility for the two refugees from Austria whom they had taken into their home. Walter's German Jewish Aid Committee Card stated that he was a scholar with the guarantor Mr C. Hauff.[44] After Walter's train on 8 August there were just two subsequent Kindertransports from Vienna.[27,44] With war starting on 3 September 1939 Walter was extremely fortunate to be able to leave Vienna to go to England. If his transport had been arranged for just one month later he would have been trapped in Vienna like his parents. In total, 2,844 children travelled from Vienna to England on the Kindertransport.[45]

In many interviews after the announcement of his Nobel Prize, Walter emphasised how grateful he was to the Kindertransport scheme which saved his life. He said:

> The British don't get the immense credit they deserve for this. There was a coalition that managed to get through Parliament a special law to allow Jewish children from Germany and German-occupied parts of Europe to get special travel visas.[23]

Another young man from Vienna who took the Kindertransport was Josef Eisinger who became a very close friend of Walter. He had also been a student at the Akademisches Gymnasium but was not in the same class as Walter as he was one year younger. He had not joined the Chajes Gymnasium after both he and Walter were expelled from the Akademisches in the summer of 1938 after the Anschluss. Josef (often nicknamed Terry) was to share rooms with Walter in internment in the UK and Canada, and also subsequently when they lived in Toronto. Like Walter, he was to become a distinguished physicist. He wrote in some detail about his experiences with the Kindertransport:

> Each child was allowed to take one small suitcase, and I recall lengthy discussions regarding the selection of items that would be most useful to me in the Great Unknown. Papa's riding boots from World War I were among them, as was a wooden "mushroom" used for darning holes in socks, and 1 was indeed to make use of both of these items. The emphasis on darning socks sounds strange today, but before the invention of synthetic fibers the heels and toes of socks were forever wearing out. Mutti [mother] gave me a refresher course in darning socks, sewing on buttons and doing laundry by hand, and I practiced what was considered to be of paramount importance in England: how to use a knife and fork...

When the day of my departure arrived ... the station was crawling with police and Gestapo men, who searched our suitcases for money and other contraband. They taunted the unhappy parents and forbade them to display any emotion or distress while saying good-bye on the platform — on pain of the transport being cancelled.[46]

Josef also wrote about the train journey and his unexpectantly adventurous arrival in England:

All I can recall of the journey is looking out the window of our sealed carriage and watching the familiar Alpine landscape fade into the novel and flat countryside of Holland. In Hoek van Holland our train was shunted onto a railway ferry and we crossed the Channel still ensconced in our carriages. When we finally arrived at London's Victoria Station, nametags were hung around our necks and as we sat on wooden benches in a cavernous waiting room, our names were called out, one by one ...

That did not diminish my delight to find myself in the strange and wondrous city of London, so different from Vienna. Here people put milk in the tea instead of lemon, ride in double-decker buses that drive on the wrong (left) side of the street and traveled in underground trains with upholstered seats instead of wooden benches.[46]

Walter had registered with the local police in Sussex on 23 August 1939.[44] Then, in September, just as the Second World War was starting, he chose to train as a farm labourer in Sittingbourne, Kent. This scheme offered an experience of farm work to young men who were aged 13 or older. The period of training was eight weeks and the boys learnt to plough, tend and

harvest crops. They were taught to drive a tractor, milk cows and groom horses. With their training complete they were assumed to be qualified to have paid employment on a farm. Given his serious academic background up to this point, Walter had, somewhat surprisingly, decided to take this agricultural route. Walter recalled he received no pay for the work but he was given a treat of a chocolate bar each week.[6] He also helped to dig air-raid shelters for people on the farm. He explained that in Vienna he had "seen too many unemployed intellectuals during the 1930s" and he was thinking about his future. He subsequently said:

> My own intention at that point, in spite of my interest in science, was I felt I'd better do something where I could become really independent. I wanted to become a farmer. So it was arranged for me to go and work on a training farm where, presumably after some period, I would come out of there. It was not exactly an agricultural college, it was a training farm. But nevertheless, I learned, so I pulled carrots and looked after the piglets and I did all those things. Well then, in England, I, unfortunately, on that farm, contracted what turned out to be meningitis, so I was very ill. Many of the drugs that we now have didn't exist, but sulphur drugs had just been invented. So I pulled through. It was touch and go. After that, I was very weak. Of course, as a young person, you don't really understand how that is. You're no longer ill, but just weak. In any case, going back to the farm was out of the question.[1]

He recalled that his meningitis started with an excruciating headache. The caring Mrs Hauff then arranged for him to have a bed at an isolation hospital. A kind nurse (who Walter noticed was highly attractive) even found some Swiss newspapers to enable him to read the latest news in German — his English was still poor at that time.[6] He came close to death but pulled through.

Fig. 14. East Grinstead County Grammar School. (Courtesy of Imberhorne School.)

After a discussion with Charles Hauff, Walter was able to enrol at the East Grinstead County Grammar School (see Figure 14) and, following his recuperation, he started there in January 1940.[6] This was 8 miles from the Hauff's house in Copthorne and took about 30 minutes by bus. He studied mathematics, physics and chemistry — a standard combination of advanced subjects taken by 16-year-olds at good English schools at that time and which are still taken by many "sixth-form" students in England today.

The school was originally called the East Grinstead Free Grammar School and was founded in 1708. It was then stated that "All the boys into this school are taught the English Language, Reading, Writing, Arithmetic, Psalmody, the Catechism of the Church of England and the Holy Scriptures."[47] In 1887 the Charity Commissioners had the insight to add "scientific, technical or literary subjects" to the curriculum. By the late 1930s the school had become selective for boys and girls and a rigorous examination normally had to be passed for entry. There was a smart school uniform. Half the places were free and fees had to be paid for the other half depending on the income of parents or guardians.[47] After the start of the war, the school had become

Fig. 15. Mr Thomas Scott, Headmaster of East Grinstead County Grammar School. (Courtesy of Imberhorne School.)

very crowded due to children being evacuated from London. In 1940, some bombs even dropped near to the school which was close to a typical flight path from Germany to London.

It is clear that the school recognised the academic potential of this unusual student from Vienna even if he did not have the skills in English to take the entrance examination. Mr Thomas Scott (see Figure 15), the kindly headmaster, started Walter in the competitive sixth form and informed the teachers to overlook Walter's inexperience with English.[6] He also set up an unconventional class in which Walter was taught English on his own for half the time and then he helped to refine the German language of his tutor. The teaching of science was quite advanced and, for example, required the use of calculus for understanding the properties of matter and heat flow, an ideal challenge for a student of Walter's ability, interest and background. Walter was only at the East Grinstead school for five months but he often commented that this brief educational experience was important for him. In his career, his education had several short and unconventional episodes which all proved beneficial for his learning and for the development of his resilient character.

However, the dark clouds of war were looming and in May 1940 the sudden attack of the German armed forces on several Western European countries was a devastating blow. Following the withdrawal of the British troops from the continent and the miracle of Dunkirk the atmosphere changed. With an invasion of England suddenly a possibility, the rapidly changing government of the UK took up extreme measures. The Government of Neville Chamberlain fell on 9 May and he was replaced as Prime Minister by Winston Churchill.

As a young man Churchill had experienced an extraordinarily adventurous life. He was commissioned in the 4th Queen's Own Hussars and saw military action in Cuba, India, the North West Frontier, Sudan, the Boer War in South Africa and the Western Front in World War I.[48] He had a no-nonsense attitude to war and this extended to the thousands of Jewish people who had made Britain their home in the 1930s. With the fall of France in June 1940 he was determined the same would not happen to the UK. With Benito Mussolini announcing on 10 June 1940 that Italy was at war with Britain, Churchill became even more concerned about nationals from the countries at war with Britain who were living in the UK. A number of Italian nationals had come to Britain in the preceding years and it was known that several had attended pro-Mussolini rallies in cities throughout the UK. With a significant population of German and Italian "enemy aliens" he was worried many would support the Nazis in an attempted invasion of Britain — and this had just happened in countries such as Norway and Holland. Accordingly, in July 1940, Churchill's patience finally ran out and he announced that Britain should "collar the lot".[49] This was a phrase that Walter often used, with some humour, to describe the internment order.

At the start of the war, all males in the UK aged between 16 and 60 with passports from countries at war with the UK had to register as "enemy aliens". They were then placed in one of three categories: A, a high-security risk and held in a special camp, B, a lower risk who were watched and also had their travel restricted, and C, with no risk at all. Initially Walter and most other Jewish immigrants were placed in Category C. This all changed in 1940 and Churchill would take no risks. Accordingly, less than one year after he had been in England, Walter, Josef Eisinger and some other young men who had come on the Kindertransport and who were now aged over 15 were interned. The internment was indiscriminate and included over 500 Jewish refugees who had fled from the Nazis to come to the UK and had been assisted by agencies who had helped them to find appropriate placings for academic research.[20] Even Otto Frisch, who had come from Germany and was making highly significant scientific contributions to the war effort, was interned. As many as eight physicists at the University of Bristol were interned: Walter Heitler and his brother Hans, Herbert Fröhlich, Kurt Hoselitz, Philipp Gross and Heinz London, and two of their students Robert Sack and Geoffrey Eichholz.[50] A major effort was at once initiated by bodies such as the Royal Society and the Society for the Protection of Science and Learning (SPSL) to free these individuals from internment camps.[20] Walter, however, was yet to be an eminent scientist and did not receive such attention. He subsequently commented:

> By civilized, peace-time standards our internment was outrageous. But most of us understood the extraordinary war-time pressures which existed at the time and felt very lucky to have escaped from the Nazi machine. Nevertheless, being interned by the "friendly" British just months after getting out [of Vienna] was a *very* bitter pill.[51]

Chapter Three

Internment

Great Britain has received justified praise for the Kindertransport. However, the same cannot be said about the decision in 1940 to intern men between the ages 16 and 70 who were nationals of the German Reich. By the summer of 1939 this included Germany, Austria and Czechoslovakia, and many of these people had been forced out of their home countries by the Nazis. The same internment rules also applied to Italian nationals after that country entered the war.

Britain had considerable experience of internment which had been applied in the First World War. Right at the start of that war it was not imposed in major numbers but after the sinking in May 1915 by a German torpedo of the *Lusitania*, a British-registered ship which had departed for New York, the reaction against Germany by the general public forced the government to set up a comprehensive internment scheme. Many German-owned shops, still open in Britain, were attacked and the government became concerned about the safety of German nationals. Initially, those interned were kept in various camps on the mainland but, with the numbers rising to over 16,000, a different scheme was required. Accordingly, it was decided to create a large internment camp on the Isle of Man.[49] This small island is situated between Britain and Ireland and is a self-governing British Crown dependency. The British Government was (and is) responsible for the defence of the island and its secure location was thought to make it an ideal place for keeping the German nationals out of the way. Consequently, when internment was established again some 25 years later in 1940, the obvious place was the Isle of Man.

Walter was in the attic in the Border Cottage at Copthorne in May 1940 when Mrs Hauff came to the door. He recalled:

Mrs Hauff said, "Walter I have some really unfortunate news for you. Two gentleman came to the house about ten minutes ago from Scotland Yard. I am terribly sorry but you have to go with them." I was close to tears as I had been very lucky to be looked after by the Hauffs and was very happy in school. She said this was happening all over the country. So I packed a small bag and went down from the attic to the main house. I had no hope and knowledge of the future.

In May 1940, shortly after I had turned 17, and while the German army swept through Western Europe and Britain girded for a possible German air-assault, Churchill had ordered most male "enemy aliens" (i.e., holders of enemy passports, like myself) to be interned.[6]

Walter's subsequent friend Josef Eisinger also illustrated the improvised aspects of internment:

I had just turned sixteen and was working in the kitchen of the Park Royal Hotel in Brighton ... In view of my questionable prospects at the hotel, it won't be difficult to understand why I was not greatly distressed when one day, two policemen looked me up in my pantry and asked me politely to accompany them to the police station. I assumed that I was being summoned to fill out another registration form related to my status as an enemy alien. I dried my hands, and left the sink full of dishes to join them, when one of the policemen suggested that I might want to bring a toothbrush along. The date was 16 May 1940, shortly after the Dunkirk evacuation, and it finally dawned on me that my life was about to take another sharp turn... At the police station I found myself among a dozen or two of other "enemy aliens" who had been rounded up that day, almost all of them Jewish refugees.

Later in the evening, a police van took us to the Brighton race-course, the first of several improvised internment camps that I was to inhabit in the next twenty months. We were brought into a large hall with a cement floor and were issued cotton pallets which we filled with straw that had been piled high in a corner of the hall. When stuffed with just the right amount of straw, the pallets made acceptable bedsteads.[46]

Both Walter and Josef were sent by train to Huyton, a suburb of Liverpool. Several details of Josef's writings on their internment also applied to Walter:

There are good reasons why internment caused me little anguish in the beginning in spite of being deprived of the limited liberty and privacy I had enjoyed at the hotel. I was suddenly freed from the drudgery of my previous jobs and most important, I found myself in the company of boys of my own age and background for the first time since leaving Vienna. And beyond that, although I did not know it, internment was to open new vistas for me, vistas that would shape the rest of my life.

From the Brighton racetrack we were shipped by train to another camp, an unfinished housing development in Huyton, a suburb of Liverpool. Since there were not enough habitable houses to hold all inmates, the youngest ones, myself included, had to sleep on the ground in pup tents. More distressing was the diminutive food allowance. Our daily ration was a single, inch thick slice of white bread, occasionally supplemented by a spoonful of jam or a salt herring fished out of a wooden barrel. For the first time in my life I experienced hunger for days on end. While attempting to trisect my daily bread ration so I would have a slice of bread for three 'meals' a day, I sliced off a piece of my left thumb and I still retain the scar as a reminder. It was in Huyton that I made contact with Walter Kohn who, although he was a year older, had been a fellow student at the Akademisches Gymnasium. We quickly formed a friendship that lasted for more than seven decades.[46]

After one month in Huyton, Walter and Josef were marched to the Liverpool docks. There Walter recalls being spat at by local women who thought they were Nazis.[6] They boarded a ship that took them across the Irish Sea to the seaside resort of Douglas in the Isle of Man. The camp there was a collection of small hotels that had rapidly been requisitioned. Josef and Walter shared a room with another friend Rudi Cohen.[46] The internees formed various disparate groups including a group of German seamen who had mutinied and turned their boat over to the British, and veterans of the Spanish Civil War. Josef wrote that the conditions were tolerable and they were given rhubarb which was readily available on the island. They were even allowed to swim on the beach, watched by genial guards. Walter recalled that the food rations were miniscule and he lost 30 lbs in weight.[6] Walter subsequently found out that the soldiers who were guarding the internees on the Isle of Man were stealing the food meant for them and selling it on the black market in the UK as there was a major food shortage thanks to the Atlantic blockade by U-boats.[6]

Charles and Eva Hauff could not understand why Walter, who had to escape from the Nazis, was interned. They arranged for some physics and mathematics books to be sent to him (see Figure 16), fully expecting Walter to return soon to Sussex to continue his studies. Walter kept these well-thumbed books for his whole life as they kept up his interest in physics and enabled him to study quite intensively during many dull moments. They included "Properties of Matter" by D. N. Shorthose and "Heat" by R. G. Mitton.[52,53] These books required an understanding of calculus and were at the level of first-year undergraduates.[54] At the camp, lectures were given by inmates and they included some in mathematics and physics. Improvised concerts, so central to life in musical Vienna, were also performed.[46]

Fig. 16. Books sent to Walter Kohn in the Isle of Man (Shorthose and Mitton) and the Ripples Camp (Hardy). (Courtesy of Nate D. Sanders Auctions.)

The number of interned "enemy aliens" on the Isle of Man was getting out of control and the British Government appealed to the Commonwealth countries and Dominions to find space for some of them. To implement this plan, it was necessary to take a dangerous trip across the oceans. One boat, the *Dunera*, even took internees to Australia under horrendous conditions.[20] The most famous ship involved was the *Arandora Star*. In July 1940 this ship was heading for Canada unescorted after leaving Liverpool with over 1,000 internees from Italy and Germany, including many Jewish men. When quite close to the Irish coast the boat was hit by a torpedo from a U-boat. Over 800 men drowned.

Uberto Limentani was a distinguished Italian scholar who was a colleague of the author at Magdalene College, Cambridge in the 1980s. In 1940 he was a Jewish refugee and was on the *Arandora Star* after being interned a few weeks previously. He told me he was in the water for several hours but was one of the lucky ones to be saved. He managed in due course to get back to the mainland where he gave frequent broadcasts in Italian on the BBC Radio London programmes. When he travelled in Italy after the war strangers who talked to him would frequently say "Where have I met you before?".

Being amongst the youngest internees on the Isle of Man, Walter and Josef expected to be among the first to leave. Walter recalled an announcement made by the Camp Commander:

We expect that a substantial number of you will be sent to Canada and the remainder will be returned to the mainland to other camps. You will be allowed to make a decision as to which option you prefer. You must make your decision today.[6]

The older internees who had jobs or family members in England preferred the option to return to the British mainland. Walter, however, subsequently commented that the food had been so meagre in the British camps that he chose Canada. He recalled also that he had already thought about emigrating to Canada when he was in Vienna and had applied unsuccessfully there for a visa.[6]

Josef Eisinger wrote that Canada brought up thoughts of vast forests, trappers travelling on canoes on its rivers and lakes, and the best hockey team in the world.[46] The Polish Passenger ship, the *Sobieski*, was chosen to be the boat that would take these bright young men from Vienna to Canada from the port of Glasgow (Figure 17). Walter recalled having his first good meal for many weeks in Glasgow of "canned cured beef".[6]

The ship left Scotland on 5 July 1940 bound initially for Quebec.[46] The boat was so crowded that it was only possible to sleep on the floor or in bunk beds via several shifts. There were over 1,500 people on board including 500 German prisoners of war who were separated from the internees and others by barbed wire. There were also some British children who were being evacuated from their families. The ship was part of a convey of merchant ships which were protected by destroyers. However, halfway through the voyage one of the propellors of the *Sobieski* failed and it fell behind the convoy. Depth charges had to be released in case a U-boat was close by. Walter subsequently heard that a U-boat had been hit by one of the depth charges.[6]

The intellectual promise of the Jewish internees on the *Sobieski* was extraordinary.[55] Like Walter Kohn and Josef Eisinger, many went on to remarkable academic careers in North America after the war. Those who became scientists included Carl Amberg from Aachen who became Professor

Fig. 17.   The *Sobieski*, ~1940. (Courtesy of Kevin Blair.)

of Chemistry and Dean of Graduate Studies at Carleton University. Alfred Bader was from Vienna like Walter but attended the Real Gymnasium. He was to obtain his PhD in chemistry at Harvard and founded Aldrich, the world's largest supplier of research chemicals. He also became a major philanthropist for his alma mater of Queen's University, Canada and several chemical societies. Vernon Brooks had a career as a distinguished neuroscientist working at McGill, Rockefeller and Western Ontario universities. Ernst Deutsch became a geophysics professor and expert on plate tectonics. Ernest Eliel was made President of the American Chemical Society and did leading research on the stereochemistry of organic molecules. Josef Kates was to be a celebrated computer pioneer who designed the first digital game-playing machine. He had attended the Goethe Gymnasium in Vienna. He escaped from Austria to Italy one day before the start of the war and, despite enlisting in the British army, he was still interned and transported to Canada. Hans Reichenfeld had been in the same class as Walter at the Akademisches Gymnasium and had also escaped from Vienna on the Kindertransport. He would have a distinguished medical career specialising in geriatric medicine.

Also on the boat were the physical chemist Kurt Guggenheimer and the pure mathematician Fritz Rothberger who had been working in Cambridge after fleeing Germany. They were to become important teachers of Walter in Canada. In addition, there were several similar examples of future outstanding academics in the humanities and social sciences on the *Sobieski*. This was the last boat that took civilian internees to Canada.

The *Sobieski* arrived at St John, Newfoundland on 15 July 1940 after a journey of ten days and then sailed up the St Lawrence to Wolfe's Cove, a mile up river from Quebec City.[46] A ravenous Walter recalled being given another good meal of boiled eggs and ham — with some of his hungry Jewish colleagues forgetting their Kosher rules.[6] A train took the weary travellers to Trois-Rivières where the camp was next to a baseball field. They were marched through the streets where the people were expecting Nazi soldiers and were surprised to find a motley crew. Their camp was primitive with most sleeping in a large exhibition hall with only one working lavatory.

Walter and Josef were ordered to serve food supplied by the camp kitchen but the rations in Canada were much superior to what had been supplied before.[46] Even showers were available — something that had hardly been established in Britain. The camp had some German prisoners of war and fights broke out between Nazis and anti-Nazis. A barbed wire fence was erected to keep the two groups apart. This was their "home" for a month. Then, with over other 700 internees, they were transported by train to a camp near Fredericton in New Brunswick, which was six hundred kilometres to the east of Quebec. The internment camps in Canada were labelled by the alphabet and this was designated Camp B, and was also known as Ripples Camp.[51] It was in a heavily wooded forest. The camp was initially quite primitive but the internees took to carpentry to improve the facilities.

The Ripples Camp was part of a huge organisation of over 40 internment camps in Canada, nearly all in New Brunswick, Alberta, Quebec and Ontario (see Figure 18). The guards in the camp were expecting committed Nazis and were surprised to find Jewish men who were strongly opposed to Hitler's regime. Nevertheless, there were the usual arrangements in the style of a prisoner-of-war camp with armed guards, machine gun posts, floodlights, watchtowers and barbed wire fences. The Red Cross visited to check on the facilities but the internees refused to acknowledge them saying they were not prisoners of war. The camp was so remote that escape was very unlikely. There was deep snow in the winter but no shortage of wood, which was cut down in the forest by the internees to keep warm. The internees found the food to be quite generous compared to what they had received before and this sometimes even included maple syrup.[51]

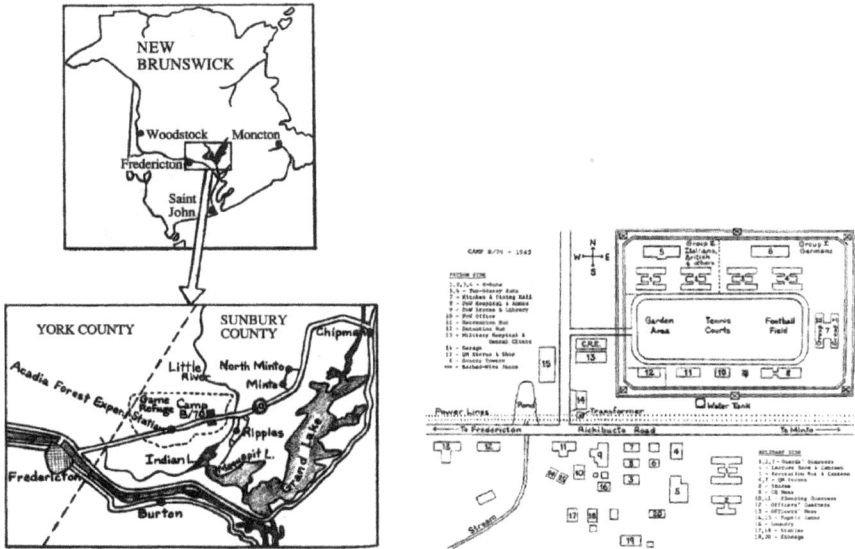

Fig. 18. Camp B, the Ripples Camp in New Brunswick, Canada. (Courtesy of the Government of Canada. Reproduced with the permission of Library and Archives Canada (2024).)

The orthodox Jews in the camp, numbering close to 200, did their best to keep to their strict Jewish customs and diet. They also insisted on observing the Jewish Shabbat and holidays. Although proud of his Jewish heritage, Walter was not part of this group. He always considered himself to be a "reformed Jew" in his own words, retaining a Jewish identity but not following the orthodox route of his grandparents.[6] There was a hunger strike from the orthodox internees requesting Kosher food and this was eventually supplied by January 1941 thanks to the help of the Canadian Jewish Congress.[51] The internees also organised themselves to press for their own release. However, with the Battle of Britain raging in the summer of 1940 they were almost forgotten for a time and were not a priority.

The Canadians were very surprised to find that many of those interned in the camps were well-educated, English-speaking and bitterly opposed to the Nazis. This was even raised in the Canadian House of Commons.[56] Canada was to benefit significantly in due course as several of those interned, including Walter Kohn, stayed on to be educated in universities in Canada and as many as six were eventually awarded the Order of Canada, contributing significantly to Canadian life in a broad range of areas including music, art, literature and science.

Fig. 19. Six friends in the Ripples Camp. Back row, left to right: Walter Michel, Josef Eisinger, Walter Kohn. Front row, left to right: Walter Odze, George Sanger, Josef Weininger. Note the similarity between Josef Eisinger and Walter Kohn. ~1941. (Courtesy of Josef Eisinger.)

There was plenty of spare time at the Ripples Camp and this gave Walter an opportunity to teach some physics to his friend Josef (see Figure 19).[46] There is no better way to learn a subject than by teaching and this was a skill that Walter put to good effect throughout his academic career. A key person in the Camp was Alfons Rosenberg, a schoolteacher from Berlin.[51] He noticed that the majority of the internees in the camp were aged under 20 with several having taken the Kindertransport. Some older internees had been studying or undertaking research at British universities and they formed a group to give instruction on a broad range of subjects across the sciences, social sciences and humanities. Appeals brought in texts and exercise books and Rosenberg was even able to arrange for the best students to sit for nationwide McGill University examinations organised for entry to a university in Canada.

Present at the camp were some physics experts who had already published scientific papers. This included Kurt Guggenheimer who had studied under Fritz Haber in Berlin and had published two significant articles on the binding energies of nucleons.[57,58] This influenced Walter Elsasser who was to develop the theory further and was to become a close colleague of Walter Kohn's in San Diego in the 1960s. On the infamous Kristallnacht of 9–10 November 1938, Guggenheimer was imprisoned in the Dachau concentration camp. Following support from the SPSL he was allowed to leave Germany on 31 August 1939, just three days before the start of the war.[59] He was able to work at Cambridge University with the surface scientist Eric Rideal. Another subsequent paper from Guggenheimer on nuclear energy levels, initiated in Cambridge, was to be communicated by Paul Dirac to the *Proceedings of the Royal Society* in 1942.[60] It applied ideas adapted from chemical bonding to nuclear structure. Dirac acted as a referee for Guggenheimer and wrote:

> I have looked through the theoretical parts of Guggenheimer's papers. The work is of a qualitative nature and much of it is rather rough, but it does contain some good ideas. Although the data available to work with are very meagre, Guggenheimer has studied them very carefully and his work gives me the impression that he is, at bottom, a sound researcher having the necessary physical insight to be able to obtain useful results.[59]

For any other physicist this would have been a lukewarm reference but from Dirac this was a strong recommendation. Guggenheimer's paper had a major influence on the Nobel Prize-winning nuclear theory of Maria Goeppert Mayer (who also would become a colleague of Walter's in San Diego).

At the Ripples Camp, Walter attended Guggenheimer's daily classes in physics. This allowed him to hear about quantum mechanics, probably for the first time, on which he would, in due course, devote his research career. However, the opportunity only lasted a few months as Guggenheimer was released from internment in January 1941 after strong representation from Cambridge University and from Lawrence Bragg whose research group he joined briefly.[59] In the late 1940s Guggenheimer became a lecturer at the University of Glasgow and became a naturalised British citizen.

Walter also studied mathematics with Fritz Rothberger (see Figure 20) who was, in the words of Walter, "the son of the owner of the most elegant men's store in Vienna".[6] Rothberger had obtained a doctorate at the University of Vienna and studied in Kraków. He became a pioneer in the development

Fig. 20. Fritz Rothberger. (Courtesy of the Canadian Mathematical Society.)

of combinatorial set theory and a topological space is named after him. Straight after the Anschluss, the Nazis confiscated the property of his family including very valuable art and ceramic collections. Fritz Rothberger's escape to England was assisted by the SPSL.[61] He applied in April 1939, just after the Anschluss, and this was a very late time to get assistance from the SPSL. However, he held significant stock in British banks and this financial security was sufficient for him to get the necessary permits to come to the UK where he was expected to be self-sufficient. During this period, there were many other applications to the SPSL from distinguished mathematicians from Germany, Austria and Czechoslovakia. Several of them could not be assisted as they did not have the backing of significant financial resources and some were to die in the Holocaust.[20]

In England, Rothberger was not formally associated with a university and stayed with friends in numerous places including Dorset, Oxford and Cambridge. In May 1940, following the German Blitzkrieg, he was interned like many others at the same time as Walter. He had managed to write a paper which was published in the *Proceedings of the Cambridge Philosophical Society* in French and communicated by the distinguished mathematician Abram Besicovitch of Trinity College, Cambridge (who was to become the tutor of John Pople).[62]

Walter's opinion on Rothberger was very positive:

> Rothberger normally taught us out of doors where he wore
> shorts and boots and nothing else. He used a stick and a sandy
> area for a blackboard to teach us about the different types of
> infinities… He was a most kind and unassuming man whose
> love for the intrinsic depth and beauty of mathematics was
> gradually absorbed by his students.[54]

Walter's subsequent papers in theoretical physics show a purity of
thinking that links back to the fundamental mathematical training he
received from Rothberger. Walter also stated that there were just two in
Rothberger's class.[11] The other was Jim Lambek who would eventually
become Professor of Pure Mathematics at McGill University. In an obituary
of Rothberger, Lambek wrote: "I can safely say that, had it not been for Fritz
Rothberger's teaching and friendship at that time, I would never have become
a mathematician or entered academia."[63,64]

Walter's education at the Ripples Camp, although unconventional, was of a
very high quality in physics and mathematics. By working as a lumberjack for
20 cents a day, which he enjoyed, he also managed to save up enough money
to buy the books *Introduction to Chemical Physics* by J. C. Slater, and *A Course
of Pure Mathematics* by G. H. Hardy (see Figure 16).[11,65,66] Little did Walter
know that the subject of the book by Slater, Chemical Physics, would win him
the Nobel Prize over 50 years later and, indeed, Slater developed a quantum
mechanical procedure dubbed the $X\alpha$ method that had a close relationship to
Walter's DFT. Years later, Walter would have an academic dispute with Slater
when he found an error in one of Slater's papers. These books would have been
very challenging for a 17-year-old but Walter was exceptional.

Walter enjoyed the forestry labour in the camp as it kept him warm in the
winter whereas "the watching guards were always freezing cold." He also was set
to work sewing camouflage nets for military use.[6] Walter, however, commented
that he always felt imprisoned in all the camps he was sent to and he found
it depressing that he still had no control over his life.[6] He wore the standard
camp uniform of denim clothes, a great coat and heavy boots. There were still
armed guards and the internees had to wear a large red circular sign on their

backs which would make them easy targets if they tried to escape, which very rarely happened. Walter remarked that "Canada was a huge country, and even if someone escaped it was unclear where they could go."[6]

Musical concerts were also arranged in the Ripples Camp. Mozart sonatas were performed by Hans (John) Newmark (Neumark) who went on to a highly successful career as a concert pianist in Canada. The camp canteen was surprisingly well stocked with cigarettes (in addition to the extraordinary ration of 50 per internee per week), fresh fruit and useful items such as razor blades. Regular deliveries from Fredericton even included regional delicacies such as clams, lobster and pork which, although strictly non-Kosher, proved to be very popular.[51]

At the end of December 1940 there was a special visitor from Britain, Alexander Paterson. He wrote a report for the British Home Office on how best the internees could assist with the war effort and recommended their release in Canada.[51] However, this was opposed by the authorities in Ottawa who recommended that if they were to be released, it should be to the UK or the USA. Over 500 internees from across all the Canadian camps were returned to the UK in this way to continue their occupations there.[51] Walter and Josef were too young to be able to join this group.

Max Perutz was also born in Austria and, before the war, was studying for a PhD in crystallography at the Cavendish Laboratory at Cambridge University. Like Walter, he was interned briefly at Huyton and the Isle of Man. He set sail from Glasgow to Canada on the *Ettrick* in July 1940 in the same month as the *Sobieski*. The conditions in the ten days on the *Ettrick* were appalling and much worse than those in the *Sobieski*. Perutz's time in internment in Camp L overlooking the St Lawrence River was short but he recalls some excellent lectures in physics given in the camp by the future atom bomb spy Klaus Fuchs.[67] However, Max Born made a strong and successful case that was sent on by the SPSL to the Home Office for Fuchs to be released from internment, stating that he was "capable of undertaking work of immediate national importance".[68] Born even asked Einstein and John von Neumann at Princeton to send books to Fuchs.[69]

With scientists such as Lawrence Bragg making a strong case for him back in England, Perutz was also informed in November 1940 that he was going to be released.[67] He then had to make the long, 1,000 kilometres train trip east to Nova Scotia to take a boat back to the UK. The trainline went via

Fredericton and he was interned in the Ripples Camp for the cold month of December 1940. Perutz subsequently wrote about the camp:

> My release order arrived and I was transferred to another camp… This is a purely Jewish camp, in a way the best I have seen. However, I am sick of waiting and fed up to the brim. I just can't understand how a man worth a 10 years imprisonment term does not commit suicide… I am so tired of it all and so homesick that I nearly cry when I think of it.[67]

It seems that Perutz and Walter did not meet in that cold month in the Ripples Camp. In 2001, Walter told the author of this book that he did not recall ever meeting Perutz, one of the three Austrians who left the country of their birth in the 1930s and subsequently were awarded the Nobel Prize for Chemistry (the others being Walter Kohn and Martin Karplus).

By the summer of 1941, more internees were being regularly released back to the UK for war work and the future was looking more promising for Walter. The internees were allowed to indicate whether they would prefer to return to the UK or stay in Canada. Apart from the Hauff family, Walter did not have links with the UK and he was not yet known in any of the universities. He had enjoyed the unconventional education he had already received in Canada and his preference was to stay in that country and continue his education there.[6]

On 19 June 1941, the remaining 462 refugee internees were moved out of the Ripples Camp to make room for the increasing number of German prisoners of war being sent to Canada. After the war, the contributions made by several of the Ripples camp internees to the Canadian nation were remarkable. Six were awarded the Order of Canada: Walter Homberger (Managing Director of the Toronto Symphony Orchestra), Helmut Kallmann (distinguished musicologist), Henry Kreisel (novelist), John Newmark (concert pianist), Erwin Schild (rabbi) and Max Stern (art dealer).[51]

Walter and Josef, and many of their age group, were moved to Camp A at Farnham, 60 kilometres southeast of Montreal. It was anticipated that this move, closer to the main Canadian population, would eventually facilitate their release. The school at Camp A was run by William Heckscher, who was a non-Jewish German art historian.[54] Walter commented very favourably on the school of Heckscher, whom he also met after the war at the Institute for Advanced Study in Princeton.

In 1941 the USA had not yet entered the war and the outcome was still uncertain. However, Walter felt he had an obligation to his parents to continue to study as hard as he could, despite the challenging and unconventional circumstances. In the camps he did have the opportunity to read Canadian newspapers but he had not read anything yet about the "Final Solution" of the Nazis. In 1941, he continued to receive occasional letters via the Red Cross from his sister Minna, who had not been interned in England, and his parents, who were still in Vienna.[6]

William Heckscher managed to arrange for his star internees to travel to Montreal in September 1941 to take the Senior Matriculation Examination of McGill University. Walter obtained very good grades in the subjects he took — algebra, geometry, trigonometry, and physics.[54] He even obtained a good mark in chemistry — his first and last examination in that subject which would become so important for him over 50 years later.

Walter later commented that "Scotland Yard" in England had done some diligence on the Jewish internees in Canada, including himself, and realised they were genuine refugees and no threat.[6] Some of the refugees had family members in Canada who could take them in but Walter had no such connections. Then on 21 October 1941, out of the blue, both Josef and Walter received identical letters from Hertha Mendel, who was the wife of Bruno Mendel, a distinguished faculty member of the University of Toronto (see Figure 21). Her letter stated:

Dear Mr Kohn and dear Mr Eisinger

I hope you don't mind that we are writing to you both together. We have heard from Charles Kahn that you are great friends and we want to tell you that we would like to be your sponsors and are hoping to have you with us very soon.

I suppose you would like to know who we are. We are German refugees who left Germany in 1933 and are now

Canadian. My husband is a Physiologist and Professor at the University of Toronto. We have three children of about your age and we are all looking forward to seeing you soon.

    With best wishes, Yours sincerely, Bruno and Hertha Mendel.[46]

    Bruno Mendel had a position in the Banting Institute, named after Frederick Banting who discovered insulin together with Charles Best. After German military service in World War I, when he was seriously wounded, Mendel had worked in a German teaching hospital and developed his own research programme in Berlin, collaborating with Otto Warburg who won the 1931 Nobel Prize for Physiology or Medicine for his discoveries on respiratory enzymes. Like many others from Jewish families, Mendel left Germany in 1933. He first went to the Netherlands and after that became an Assistant Professor to the Banting Institute in 1936. There Mendel studied cholinesterases, enzymes which are important in the nervous system. For this research, and as a Canadian citizen which was part of the British Commonwealth, he would be elected a Fellow of the Royal Society in 1957.[70]

Fig. 21. Bruno and Hertha Mendel. (Courtesy of the Royal Society (Bruno Mendel) and Josef Eisinger (Hertha Mendel).)

In the early 1900s, Hertha's mother Toni, who was also Bruno Mendel's aunt, had been a very close companion of Einstein.[71] It seems that a previous internee who had already been released, Charles Kahn, had told the Mendels that Walter and Josef played recorder duets together.[46] This implied to Bruno and Hertha Mendel, who often performed Mozart concerts, that the pair of young men would be a good choice to join their family. The deep musical heritage of the Theobaldgasse in Vienna had come to help Walter in a subtle way.

Josef Eisinger wrote about his feelings of freedom after his release from the internment camp:

Finally, early one frosty morning in January 1942, an army truck took me through the double gates of Camp A and deposited me at the Farnham railway station. I was handed $15 (the Mendels were later charged for it), and bought a ticket to Toronto, and stood by myself — itself a strange feeling — on the platform, savoring freedom after 20 months of confinement. I was wearing an army great coat, dyed civilian dark blue, a disguise that allowed me to mingle inconspicuously with ordinary men and women — as if that were a perfectly natural thing to do. (It is a sensation that has never left me entirely)…

I never felt resentful towards Britain for her misguided internment policy in those dark, fearsome days following the fall of France… I admired Britain for, in the end, she stood up to Hitler and was the only nation willing to accept some 10,000 Jewish children who escaped from the Nazis in the Kindertransport. But then, I had been one of them…

I could hardly have hoped for a more congenial environment than that I found myself in following my release in January 1942… Since the Mendels escaped from Nazi Germany so early, they were able to take many of their possessions with them. As a result, their Toronto home retained much of the character of the sumptuous villa in Berlin-Wannsee where they had lived along with Hertha's widowed mother Toni, the villa's owner… The house was full with books, music and paintings… For Walter and me, just emerged from two years living in barracks, our new home was utopia.[46]

Walter and Josef shared an attic room at the Mendel's house at 98 Bedford Road, Toronto and they were quickly welcomed to the family. So Walter once again had found freedom in a small attic room (see Figure 22 where the attic garret is in the upper centre of the picture). Walter and Josef helped the Mendel daughters Ruth and Anita with their mathematics homework.[46] Their son Gerald was soon to be commissioned in the Canadian Army and was to be a member of the Canadian troops at D-Day and the liberation of Europe.

The Mendels also took in other children including the daughters of Francis Simon, Kathrin and Dorothee, who had been evacuated from Britain. Simon was a Jewish physicist who had left Breslau in Germany in 1933 for Oxford.[72] He had made a crucial contribution to the Allied war effort in

Fig. 22.   98 Bedford Road, Toronto. Home of the Mendel family. (Photograph by Bob Krawczyk. Courtesy of the TOBuilt database.)

developing the diffusion method for separating uranium isotopes that was to be applied on an industrial scale in the Manhattan project. Walter and Josef got to know the Simon daughters and they met Simon himself when he visited Canada to undertake secret war work.

Obtaining the release of Fritz Rothberger from internment took much longer as he could not be sponsored as a student, like Walter and Josef. Permission had to be obtained from both the British and Canadian authorities and a job for him had to be found if he was to stay in Canada.[64] The SPSL organised several supporting letters for Rothberger, including from J. H. C. Whitehead in Oxford, who wrote on 18 July 1943:

I have met Fritz Rothberger on two occasions, the International Congress at Oslo in 1936, and in 1939 (or possibly early 1940) in Oxford. I saw quite a lot of him and formed a good impression of his character, which was confirmed by my wife. Knowing his background, I would be very surprised if he was in any way disloyal to our war effort since this for him would be more like treason than patriotism. He has some good mathematical publications to his credit, and is worthy to hold a research post.[61]

Abram Besicovitch, who would soon replace John Littlewood in the prestigious Rouse Ball Chair of Mathematics at Cambridge and at that time was teaching John Pople, wrote very strongly on Rothberger's mathematical ability and personal integrity. Shortly after receiving these letters the President's Committee of the Royal Society recommended the release of Rothberger on 23 July 1943 and this was granted by the Home Office on 30 July.[61] He took up a teaching post in mathematics at Wolfville, Acadia University, Nova Scotia. He also had subsequent teaching positions at the Universities of New Brunswick and Windsor. When he retired, Rothberger returned to Wolfville as Honorary Distinguished Professor of Acadia University. He had fallen in love with the forest wilderness of Canada, which he had first experienced at the Ripples Camp.[64] After the war, Rothberger, who died in 2020 at the remarkable age of 98, kept in contact with Walter who always spoke about him in very positive terms.

Chapter Four

Toronto

B runo Mendel was acquainted with a distinguished colleague, Leopold Infeld, who was a Polish refugee in the Department of Applied Mathematics at the University of Toronto (see Figure 23). Infeld had collaborated with Max Born, temporarily in Cambridge, in 1934 on a new theory of electrodynamics and had even worked with Einstein at the Institute for Advanced Study in Princeton with whom he published a book entitled *The Evolution of Physics*. He was at the cutting-edge of the latest ideas in theoretical physics.[73] As a role model he was to have a significant influence on Walter.

Mendel invited Infeld to come to his house to talk to Walter over a cup of tea. Infeld probed Walter about his mathematical and physics background. Walter said that although his current interest was mathematics and physics he would like to study engineering as it would lead to better employment prospects. Infeld told him that the University of Toronto had an excellent programme in Physical Sciences — mathematics, physics and chemistry. He said that it was such a good course that Walter could always go on to engineering afterwards.[6]

So, following Infeld's advice, Walter tried to register for Physical Sciences at the University of Toronto but immediately got a negative response stating he had no background in subjects such as English and History which were also requirements for entry to the University.[6] Infeld then advised Walter to see Samuel Beatty, the Chair of the Mathematics Department and Dean of the Faculty of Arts and Sciences (see Figure 24). Beatty had been the first scholar to receive a PhD in Mathematics from a Canadian university. His supervisor was John Fields, who established the Fields Medal, thought by many to be the mathematics equivalent of the Nobel Prize. Beatty was highly respected in Toronto and would eventually become the Chancellor of the University. Beatty took Walter's case to the Registrar of the University, who was inflexible, stating that places should not be found for "charity cases for people with questionable backgrounds." Beatty would not give up and

Fig. 23. Leopold Infeld, 1960. (Creative Commons CC0 licence.)

Fig. 24. Samuel Beatty. (Courtesy of University of Toronto Archives, Ken Bell Associates Photography.)

explained to the whole faculty that there were six refugees, including Walter, who should be enrolled as "Special Students".[6]

However, the Chair of Chemistry, Frank Kenrick, declined this route. Kenrick was a chemist of the traditional school who had written a book

37 years previously with the title *An Elementary Laboratory Course in Chemistry*.[74] He said that his department was doing important research for the war effort that prevented any German nationals from entering his building. Kenrick's viewpoint became clearer when he invited Walter to his house to talk over the matter and Walter was surprised to be offered beer. Kenrick said, "I understand you are originally from Germany." Walter replied he was originally from Austria. Kenrick then surprised Walter by saying "I felt all along we should not be at war with Germany but with the communists. I studied in Germany and liked the people very much."[6] No doubt, Kenrick would have been astonished to learn that the young student whom he was talking to would win the Nobel Prize for Chemistry some 56 years later.

Dean Beatty referred the matter to the President of the University (Reverend Henry John Cody) but he could not help. Beatty then decided to redefine the Physical Sciences course so it could be taken without the option of chemistry.[6] So, at last, Walter could start his course at the University of Toronto in 1942. The University was in a very convenient position as the Mendel's house was just a walk of 10 minutes away. Beatty realised there were some gaps in Walter's mathematical knowledge (with Rothberger's teaching in the Ripples Camp being highly specialised) and he gave special classes to bring Walter up to speed. Some 25 years later Walter received an Honorary Degree from the University of Toronto and he was delighted to again meet with Beatty who had been so crucial for his university education. In many interviews after his Nobel Prize Walter always took the opportunity to express his gratitude to those people who went that extra mile to help him — especially Emil Nohel, Charles and Eva Hauff, Fritz Rothberger, Bruno and Hertha Mendel, and Samuel Beatty. He would remark that he "often felt like a leaf being blown off a tree" but these people then came into his life and made the difference for him through their kindness.[6]

In addition to Beatty and Infeld, there were several outstanding professors at the University of Toronto in mathematics and physics. In Pure Mathematics there was Richard Brauer, a leading expert in number theory and group theory with important theorems to his name.[75] He was born in Germany but had to leave in 1933 because of his Jewish heritage. Brauer had studied at the University of Berlin under Issai Schur and managed to get a first appointment at the University of Kentucky with the aid of the US

Emergency Committee for Displaced Scholars. He had also worked with Hermann Weyl at the Institute for Advanced Study at Princeton.

In addition, there was Harold Coxeter who made many contributions to geometry. From the Trinity College Cambridge school of mathematics, he had been another collaborator of Weyl in Princeton. He was to be elected a Fellow of the Royal Society and his original work on geometric shapes even inspired the graphic artist W. C. Escher and the architect Buckminster Fuller.[76] In Applied Mathematics, John L. Synge had the rare distinction of making original contributions to the theory of relativity. In addition, there was the fluid dynamicist Alexander Weinstein.[77] Russian born, he studied in Germany but had to leave in 1933 due to his Jewish birth. He then went to France from which he escaped to the USA in the dangerous year of 1940. His background linked him closely to Walter, on whom he was to have an early and deep influence through his course on the mathematics of spinning tops and gyroscopes.

Walter was also positive about his physics teachers in Toronto including the spectroscopist Harry Welsh, Malcolm Crawford in optics, who supervised the research of the future Nobel Laureate Arthur Schawlow, and Don Misener, who discovered the superfluid phase of helium at low temperatures.[11] The strong reputation of the physics department had been built up by John McLennan.[78] Brought up in Canada he had undertaken research at the Cavendish Laboratory with J. J. Thomson and he got to know Ernest Rutherford when he was performing pioneering experiments on radioactivity at McGill University in Canada in the early 1900s. In World War I, McLennan did important work for the British forces, first on detecting submarines and then on replacing explosive hydrogen by helium, extracted from Canadian oil wells, in balloons and air ships. These highly regarded contributions, and his excellent scientific connections, enabled McLennan to be elected a Fellow of the Royal Society in 1915 and to be awarded the Royal Medal in 1927. He was knighted in 1935, a rare distinction for a Canadian scientist, a few months before he died.

During his Headship of the Physics Department in Toronto, McLennan set up a seminar programme in modern physics that competed with any leading European university. In the 1920s and early 30s, speakers included Langmuir, Bohr, Franck, Kramers, Mulliken, Hund, Van Vleck, Fowler, Dirac and Debye.[79] McLennan even proposed Werner Heisenberg as a Foreign

Fig. 25.   University of Toronto. (J. Phillips attribution. Courtesy of Creative Commons Attribution-Share Alike 3.0 Unported licence.)

Member of the Royal Society in 1934, a nomination that took over 20 years for the election to be confirmed.[80]

The University of Toronto (see Figure 25) was traditional and followed many of the arrangements at the universities of Oxford and Cambridge. There were impressive stone buildings, quads and fine grass lawns. All students had to be members of a college. These had denominations — Trinity College for Anglicans and St Michael's for Catholics. Walter and Josef were members of the non-denominational University College. There they took broader courses in "Religious Knowledge" including the study of the Koran and Buddhism.[46] The university set a single three-hour examination at the end of each academic year rather than the method of regular short tests commonly applied in the USA. This yearly examination had to be passed to be awarded honours and be allowed to move on to the next stage.

Walter took an accelerated undergraduate programme in Toronto and finished the compulsory core course in one year instead of the normal two. He kept his lecture notes and problem sets in mathematics from his undergraduate days in Toronto for his whole career. They show a particularly thorough training in mathematics. The first lectures he received in geometry in 1942 were given by Beatty, Brauer and Coxeter and a colleague Irvine Pounder who all followed this up with more advanced lectures in 1943.[81] Walter took mathematics courses in algebra, analytical geometry, differential and integral calculus and differential equations. His earlier physics courses included the standard classical subjects of mechanics, properties of matter, dynamics, electricity, magnetism, light and acoustics.[54] In the North American tradition he also took some courses in humanities. As a Special Student he was allowed to replace French and German by English, which he was still learning.

After his first year, Walter took more advanced mathematics courses in 1943 and 1944 including partial differential equations and group theory. In physics he was then introduced to classical dynamics, quantum mechanics and even a special course in variational principles in physics, which he would decisively come back to again and again in his research career in theoretical physics.[54] With his colleague Byron Griffith, Synge had published in 1942 a classic textbook *Principles of Mechanics*.[82] Walter benefitted considerably from their lectures and referenced this book in his first papers. He also took a course in the application of physics to astronomy.

Being a year younger that Walter, Josef Eisinger initially joined a high school in Toronto for six months before starting at the university. He wrote that it was Walter's influence, in the internment camps and at the Mendel's house, that directed him to the courses in mathematics and physics.[46] Despite being born in Austria, Josef had to take the chemistry course, unlike Walter, but was similarly banned from entering the chemistry department by the difficult Professor Kenrick. However, the chemical engineering department allowed Josef to take the required chemistry practical classes which he completed working mainly nights in a deserted building.

The summer vacations allowed Walter and Josef to contribute to some interesting projects that employed scientific methods. Walter's first summer job for the Sutton-Horsley Company involved testing electrical circuits in cockpit instruments for use in military planes. In the next year he worked for Koulomzine, a mining exploration company based in Quebec.[11] Josef wrote about their summer activities in which they used "ingenious geophysical techniques to map the rock formations in gold mining claims which were mostly located in the sparsely populated bush country of northern Ontario and Quebec."[46]

Josef also wrote about a compulsory extra-curricular activity for him and Walter:

> This being wartime, all male students were enrolled in the Canadian Officers Training Corps (COTC) and, on two days a week, we wore our battle dress uniform to classes, were drilled, and had to listen to boring lectures about the army's command structure. At the end of the spring term, we spent two weeks living under canvas in a COTC training camp, and took part in marches and field exercises.[46]

After being in the Ripples Camp for a long period these often tedious military distractions to their studies were no major ordeal for Walter and Josef. They were fully aware of the war that was then raging at its height in Europe, and were anticipating being drafted to take part, even though this would take them away from their studies. With Canadian troops playing a vital role in D-Day on 6 June 1944 and the subsequent invasion of Europe, they decided that summer to volunteer for military service (see Figure 26).[11]

Fig. 26. Walter Kohn in the Canadian Army 1944–5. (Courtesy of the family of Walter Kohn.)

Walter was now 21 years old and Josef 20. Given their scientific interest they first applied to the Royal Canadian Air Force but they were rejected due to their German nationality. They were accepted, however, for the less choosy Canadian Army as Josef explained:

> We were quartered in the so-called Horse Palace in Toronto's exhibition grounds. Sleeping on straw mattrasses… We were indoctrinated in military law and discipline, and what seemed of paramount importance, how to salute properly and to whom. After two weeks we were considered ready for our "basic training" and received a 36 hour pass, but not before we were so massively vaccinated that I spent the entire leave in bed at the Mendels, nursing a high fever… Walter and I were in the same platoon and lived in an H-hut, so familiar to us from our internment days.[46]

Following this basic instruction they were put through more intensive training that was needed to prepare soldiers for the fierce war in Europe. This included long marches in deep snow and sub-zero temperatures, bayonet fighting, marksmanship, and learning how to use mortars, hand grenades, machine guns and anti-tank weapons. They had to crawl through barbed wire with machine guns firing over their heads.[46] Despite their religion, they volunteered for church parade on Sundays to avoid having to clean the latrines. They enjoyed singing lusty words of their own to the Anglican hymn tunes. Josef and Walter also tasted the beer in the "wet canteen".[46] At that age, Walter and Josef almost looked like brothers, and, since their fathers both came from the same part of Moravia, it is possible they had mutual distant relatives (see also Figure 19). This led to some amusing mistaken identities which showed the humorous side of Walter's character, as recalled by Josef:

The battalion commander had organized a boxing tournament and Walter, for reasons that utterly escape me, volunteered to take part. He found himself paired with an experienced amateur boxer and took a severe beating in the first round. He was too stubborn to quit and the bout ended when Walter's seconds threw in the towel during the second round. The next morning there was a battalion parade and while inspecting the ranks, the colonel stopped in front of me and barked "Good show, last night soldier!" Too unnerved to point out his mistake I shouted back "Thank you, Sir!"[46]

However, Walter and Josef were denied the opportunity to take part in the assault on Germany. Just 48 hours before they were expecting to be shipped overseas, they were ordered to report to the company commander who told them they had to remain behind because they "had been designated as sensitive in case of capture".[46] Accordingly, they were assigned to the far less exciting role of being trained as infantry instructors staying in Canada. For the remaining part of the war they trained platoon members in basic infantry skills.

In 1944, before starting with his army training, Walter had taken the lecture course of Alexander Weinstein on applied mathematics (see Figure 27). Weinstein was originally from Russia and was of Jewish birth.[77] He was educated first in Göttingen and then in Zurich where his doctoral thesis on tensor calculus and matrices was supervised by Hermann Weyl. In 1925, he was accepted for a faculty position at the University of Swansea in Wales but this post was withdrawn with much publicity following parochial local objections that a British mathematician should have been appointed.[83] Weinstein then went to Rome where he worked with Tullio Levi-Civita on hydrodynamics. He had some discussions about being an assistant to Einstein in Berlin but, following the coming to power of the Nazis, he moved

Fig. 27.   Alexander Weinstein. (From https://mathshistory.st-andrews.ac.uk/Biographies/Weinstein/.)

to France. He also spent some time in England at Imperial College and was supported by generous grants from the SPSL for two years even when he returned to France. He had very strong reference letters, including from Sir Horace Lamb, who many years before had been a close colleague in Cambridge of James Clerk Maxwell.[72,83] Following the invasion of France by the Germans, Weinstein managed to use his Russian birth to obtain a quota place on a boat from Lisbon to New York in October 1940.[77] Thus Weinstein, being also a Jewish refugee from Europe, had a special link with Walter.

Weinstein had published on applications of classical dynamics to problems such as plates and membranes.[84] Although he was not concerned with quantum mechanics, he had used techniques such as eigenvalue determinations, spectral analysis and variational methods which were being applied in quantum mechanical calculations on atoms, molecules and solids. Thus, even before taking advanced courses on quantum mechanics, Walter had been exposed to the fundamentals of the key mathematics that was to lead eventually to his Nobel Prize. The lectures by Weinstein had clearly piqued Walter's interest as, in the free time available during his army training, he wrote two papers connected with them. The first, published in 1945 and consisting of just two pages in *Quarterly of Applied Mathematics*, applied spherical symmetry to simplify the theory for the rotation of a gyrocompass.[85] Walter's mathematical analysis showed that "a spherical gyrocompass has an angular velocity of strictly constant magnitude relative to the earth." This work may well have been stimulated by his summer jobs with electrical devices and gold prospecting.

Walter's second paper was much more extensive. A summary was first published in January 1945 in the *Bulletin of the American Mathematical Society* and the full paper of 27 pages was published in 1946 in the *Transactions of the American Mathematical Society*.[86] The paper extended a work by Weinstein by using contour integration to set bounds on the motions of a spherical pendulum and a heavy symmetric top. The paper was submitted in November 1944 when Walter was just finishing his army training and shows a very impressive mathematical analysis and maturity for a 21-year-old with an unconventional education. Weinstein also read the paper on Walter's behalf to the meeting of the American Mathematical Society on 24 November 1944.[87] This was quite an honour for a student who had not yet graduated from a Master's course.

Josef Eisinger stated that Walter even gave a lecture to his bemused army student colleagues on this mathematical work.[46] There is an analogy to Walter's hero Erwin Schrödinger who also wrote papers during his time in the Austro-Hungarian army in World War I.[72] Weinstein's position in Toronto was not permanent and he left in 1948 to take up a professorship at the University of Maryland but his influence on Walter had been significant.[77]

On his return from the army Walter took the Master's programme at the University of Toronto. This allowed him to take more advanced courses and he turned his paper on the theory of tops into a Master's thesis. The lectures of Professor Arthur Stevenson then moved his interest to quantum mechanics. Stevenson had obtained his PhD at Trinity College Cambridge in 1928 supervised by Ralph Fowler. Fowler had also been the (nominal) supervisor of Paul Dirac, who had just published his great papers in quantum mechanics. Stevenson's first papers described calculations on the energies and spectra of atoms. He developed computational procedures for calculating quantum mechanical wavefunctions, including a generalisation of Hartree's self-consistent field method, for two-electron atoms.[88] In addition, Stevenson had published methods for calculating lower bounds to the energies of approximate quantum mechanical wavefunctions.[89]

Inspired by Stevenson's lectures and papers Walter wrote his first paper on quantum mechanics. Within this theory it is necessary to solve the Schrödinger equation $H\Psi = E\Psi$, where the operator H contains terms representing the kinetic and potential energy, E is the total energy and Ψ is the wavefunction from which the properties of the system are obtained. It is only possible to solve exactly and analytically this equation for very simple cases such as the electron in the hydrogen atom. For more complicated systems, Ψ must be obtained approximately and depends on many parameters. These parameters are often found by the variational principle which states that the total energy obtained from an approximate wavefunction is always above the exact energy. Thus the variation of the parameters in the wavefunction to minimise this approximate energy is required to get the best possible result.

Walter's paper, which was published in the *Physical Review* in 1947, was titled "Two applications of the variational method to quantum mechanics."[90] He added an extra term to H which contained a parameter and applied the variational principle. He then used this method to calculate the ground-state energy of the helium atom and also considered excited states. In the acknowledgement, Walter stated: "The author takes great pleasure in thanking Professor A. F. C. Stevenson for his kind advice and interest during the course of this investigation." This was the first of 121 papers that Walter would publish in the *Physical Review* or *Physical Review Letters*. The work described in this paper was also included in Walter's Master's thesis. Walter was to use some form of the variational method in many of the major papers he would write and its application within his DFT would lead to his Nobel Prize.

By 1946 Walter had finished his Master's thesis and had caught the bug of research in theoretical physics. In North America it had become traditional to move to another institution for a PhD and he wrote several letters to prospective supervisors and applied for scholarships. He took advice from Leopold Infeld and wrote to Eugene Wigner at Princeton and Rudolf Peierls at Birmingham in the UK. With three impressive papers under his wing before starting a PhD course, Walter was a very promising student and both Wigner and Peierls responded with offers of financial support.[54] Infeld recommended Peierls, whom he knew well.

Peierls had also left Germany to come to England and had previously worked with Arnold Sommerfeld, Werner Heisenberg, Wolfgang Pauli and Enrico Fermi.[91] In 1940 he had written a famous memorandum with Otto Frisch which suggested that only a relatively small mass of uranium-235 would be needed to produce a nuclear weapon. This helped to set in motion the extraordinary Manhattan project in the USA which would lead to the first atomic bomb. Accordingly, although Peierls had not been awarded a Nobel Prize, his scientific reputation was very high, especially amongst the many scientists who had worked on the Manhattan project. So, despite the misgivings of returning to the UK where he had been interned, and also the fact that the University of Birmingham did not have the reputation of some other UK universities such as Oxford or Cambridge, Walter wrote to Peierls to accept the offer.[54]

However, just one day later, Walter received a letter from Harvard University offering him the Arthur Lehman Fellowship which provided the funding for a PhD course.[54] Such was the standing of Harvard that Infeld told Walter to accept the offer and write a polite letter of explanation to Peierls. This Walter did at once. Not only did Harvard have a top reputation for research in physics but Infeld recommended that Walter should be supervised by Professor Julian Schwinger, who, although only five years older than Walter, had already established a reputation as one of the world's leading theoretical physicists.[11] He also was using variational methods in his work.

Walter was always very appreciative of the education he had received at the University of Toronto. Throughout his career he would keep in contact with the University and, even after his first appointments in the USA, he would write to his old professors with the hope that a suitable position might arise in Toronto for him. As discussed in Chapter 7, he also did not hold back in his comments to Samuel Beatty when he felt that his old Department of Mathematics was moving in the wrong direction.

Walter took part in adventurous sports and quite often received injuries (as described further in Chapter 11). In Austria, he had enjoyed skiing and he continued this activity in North America in the late 1940s, as shown in Figure 28.

Fig. 28.   Walter Kohn on skis, ~1946. (Courtesy of the family of Walter Kohn.)

Fig. 29.   Lois Kohn (née Adams), the first wife of Walter Kohn. (Courtesy of the family of Walter Kohn.)

Not long before he was off to Harvard, Walter's personal life took an important new turn. In Toronto he met Lois Adams, who was one year younger than him (see Figure 29). She was born in Ottawa on 12 July 1924 and grew up in the small town of Hornepayne, Ontario in a Protestant family. Like Walter, she had a keen interest in music. She was an accomplished pianist and studied at the Toronto Conservatory of Music for a year. She then trained at the University of Toronto School of Nursing.[92] When Walter moved to the USA for his graduate research they got engaged and she undertook public health work in New York City.

Chapter Five

Family and Schoolfriends

Since leaving Vienna in August 1939 at the age of 16, Walter had not seen his parents again. He had also last seen his sister Minna at the Hauff house in Sussex in England in the dark days of May 1940 when he was interned. So much had happened since then. Minna had stayed employed by Eva Hauff as a domestic servant during the war and was not interned. However, by the time Walter was to start at Harvard in 1946, she had returned to Vienna, got married and was in the process of getting back her father's postcard business. She had been in regular contact by mail with her brother and had forwarded messages to him from their parents during the war when they came through from the Red Cross and other sources.[6]

In 1946 Minna married Franz Pixner. He was a talented sculptor with a revolutionary background. He was born in Ried im Innkreis in Austria and undertook an apprenticeship in carpentry. This is a small town close to Braunau an Inn, which was Hitler's birthplace. Pixner then turned to wood carving and moved to Vienna to train in sculpture and drawing with Michael Powolny at the Vienna School of Applied Arts. Pixner was an active member of the Socialist Worker's Youth Movement and joined the Communist Party of Austria. Even before the Anschluss he was expelled from the University of Vienna, arrested and imprisoned twice.[93] In 1937 he went to Spain and joined the International Brigade, fighting on the side of the Republicans against the forces of General Franco. When operating behind enemy lines Pixner was seriously wounded by a land mine. In Spain he met the famous American writer Ernest Hemingway. There are reports that Robert Jordan, the young volunteer in Hemingway's classic book *For Whom the Bell Tolls* on the Spanish Civil War, was based on Pixner.[94] At the end of the civil war in 1939 Pixner was imprisoned in the Gurs internment camp in France where he was able to work on sculptures as shown in Figure 30.[95]

Fig. 30.   Franz Pixner, with his bust of the German communist Hans Beimler, a leader of the International Brigade in Spain. Gurs, 1939. (Courtesy of DÖW/Spanienarchiv.)

Pixner then moved to London where the Security Service MI5 kept a careful eye on his movements, opened his letters and wrote several reports about him. One such report stated:

> In October 1939 he [Pixner] approached the Military Committee of the Austrian Centre in London with the object of forming an Austrian Volunteer Corps to serve with the British Army. The Military Committee, however, had ceased to function by this date and Pixner's scheme collapsed. According to one source, Pixner was one of the best-trained members of the Communist party in England.[96]

Pixner was interned in 1940 after the German invasion of France but was released after two months due to the wounds he had received in

Spain, and the authorities in London felt his injuries were so serious he would not survive.[96,97] He had first met Minna Kohn in London, and they married after they returned to Vienna in 1946.[6] Minna was his third wife. MI5 continued to report on Pixner's activities. For example, on 1 December 1947 it was stated:

> Franz Pixner left this country about September on the pretext of returning to Austria. In Paris he is said to have got in touch with his former friends in the International Brigade. When next heard of in October 1947 he was in Marseilles, where it is believed he is still living. There is little doubt he is still active in International Brigade circles.[96]

In due course, Pixner turned back to his artistic interests and produced acclaimed sculpture works including the Memorial for the Freedom Fighters at the Atzgersdorfer Friedhof in 1954. He was awarded the City of Vienna Prize for Fine Arts in 1983. A road in Vienna (the Franz-Pixner-Weg) is named after him.

In Vienna, Minna (now Pixner) re-established the Brüder Kohn artistic postcard company created by her father and two uncles some 50 years previously.[41] In due course, she collected together their iconic postcards and created a museum on the Teinfaltstrasse. In 1959 she also initiated proceedings against the German Reich for reparations for the loss of property, investments and valuable items which were part of her father's postcard business based in Berlin.[98]

After their children had departed for England in 1939, Salomon and Gittel Kohn struggled to make ends meet and had to sell their furniture and other possessions to survive.[40] After the outbreak of the war in September 1939 they were still able to send messages to Minna in England either through friends or the International Red Cross.[41] The Red Cross messages were sent from the neutral country of Switzerland via Geneva. The maximum number of words was 24 and only one message was permitted each month. The messages were censored and had to be written very carefully with a coded language. There were attempts made to help Salomon and Gittel leave Vienna. On 23 April 1941 the journalist and writer Heinrich Glücksmann wrote to Erich Breuer, who was a member of the Board of Directors of the IKG in Vienna:

> One of my best friends, Salomon Kohn, who leads the postcard industry in Austria is a philanthropist on a wide scale whose love of humanity particularly affects poor Jews. He is thinking of leaving the country. Please, as the most influential person, support him with your expertise and willingness to help.[41(t)]

By November 1941 things had got so difficult for the Kohns that they had to leave their apartment at 7 Theobaldgasse, the place of so many fine memories for them. They first moved to the Jewish Ghetto in Vienna at Flossgasse. Then in June 1942 they were sent to the Kleine Sperlgasse assembly camp for one week. From there they managed to write to Gittel's sister Mala Rapaport, who had been able to remain in Vienna (see also Figure 1 for a picture of Mala as a young girl).[41] Salomon and Gittel Kohn were then transported from Vienna to the notorious internment camp at Theresienstadt.

The Nazis had established the Theresienstadt ghetto in November 1941 to intern Jewish people from the greater German Reich.[99] It was based in the town of Terezin in Northern Czechoslovakia, close to the border with Poland. The camp was especially set up to take prominent Jews over 60 years old including artists, scientists and musicians. The Nazis described Theresienstadt as a spa and used this as a trick to persuade people to give up their assets to the State in moving to the camp. The Nazis even stage-managed an inspection from the Red Cross to give the impression that Theresienstadt was a reasonably comfortable place to live.[100] This visit was witnessed by Salomon and Gittel Kohn.[41] In addition, the camp authorities forced an inmate who was a celebrated film director, Kurt Gerron, to make a sympathetic film of life in the camp.[101]

The living conditions in Theresienstadt were very poor with little food and healthcare, except that provided by the internees themselves. The sanitary and heating conditions were primitive and the people were forced to live together in overcrowded rooms. In the war, over 30,000 people who were interned died in Theresienstadt.[100] However, there was some kind of life and more notable internees were allowed to give lectures and even perform musical concerts. After October 1942, transports started from Theresienstadt to the death camps. More than 45,000 people were sent to Auschwitz. Several were also transported to other infamous death camps including Treblinka and Sobibór, or to the ghettos in the east such as those in Minsk and Lublin. This was done secretly and, even in 1945, many were not aware that people held in Theresienstadt had been transported to death camps.

From Theresienstadt, Gittel Kohn was able to have her letters forwarded by friends in Sweden to her daughter Minna in England. Minna also managed to arrange for packages of lentils and tins of sardines to be sent to her parents via neutral Portugal. Even from Theresienstadt, the courageous Salomon Kohn arranged for food parcels to be sent via friends to other starving families.[41]

With the Russian forces getting close to the German Reich the deportations from Theresienstadt increased dramatically in the fall of 1944. Over 18,000 people were deported to the death camps in October of that year.[100] The last preserved letter from the Kohns was sent from Theresienstadt on 26 October 1944.[41] Salomon and Gittel Kohn were then transported to Auschwitz two days later, with over 2,000 other people, on 28 October 1944. This was the very last train from Theresienstadt to Auschwitz.[102] Their transportation cards denoted them as numbers 989 and 990 on the train which was labelled 834-IV/1.[103] With over 1,800 people from this transportation they were then sent to the Birkenau gas chambers.

On this last train were several other notable people who were also murdered including the distinguished expert on nervous diseases from Prague, Otto Sittig and his wife Irma, the film director Kurt Gerron and his wife Olga, and Hedwig Eppstein, the wife of the Jewish Elder in Theresienstadt, Paul Eppstein.[20] The use of gas chambers in Auschwitz-Birkenau was stopped just five days later.[102]

Ada and Dr Willy Levy were also on this last train from Theresienstadt to Auschwitz. Ada, who survived the war, wrote in 1946:

On 28 October 1944, my husband and I were thrown into a cattle car that was too narrow to hold its passengers, and was immediately sealed… The appalling conditions on this transport, which lasted several days, led us to utter desperation and close to death. It seemed to us to be the ultimate form of suffering. Could anything be worse than this? … After this death journey into the total unknown, with no air, no water, no light, standing crammed together, sometimes among corpses, they took us off [the car] in the middle of the night. We didn't know where we were. SS men with whips received us… The women were immediately separated from the men, and though I hoped to still see my husband the next day, I tried to sneak one more look at my Willy in this terrifying atmosphere. I didn't understand that this would be the last time [that I would see him].[104]

The deportees had reached Auschwitz on 30 October:

We marched forward, the women, in pairs, and stood under a spotlight — right, left. I was directed to the left. We had to march at night on a dirt path. Trucks with our friends rolled past us. We could hardly stand on our feet for pain and exhaustion. We didn't know that those standing on the trucks were being led to their deaths. No one saw or heard from them again.[104]

With the last messages from his parents arriving late in 1944, Walter was still very busy in his army training. For some time he did not know about the fate of his parents with the chaos and confusion as the Russian armies advanced into the German Reich. In February 1945, Heinrich Himmler had arranged for over 1,000 prisoners remaining in Theresienstadt to be released in exchange for a considerable sum of money provided by Jewish organisations in Switzerland.[102]

He was hoping to use this as a bargaining tool for negotiations with the Western powers — something that was strongly disapproved of by Adolf Hitler. Then, near the end of the war, Walter received a letter at the Mendel's house from a survivor of Theresienstadt who had been asked by Walter's parents to let their son know that they had been deported to Auschwitz.[6] It took some time after that to receive further confirmation that his parents had perished once lists and databases on the Holocaust were beginning to be set up.

Walter was deeply traumatised by the murder of his parents and he spoke about this in many interviews, especially after he had won the Nobel Prize. He was particularly scathing of the murder of his Viennese cousins who were just young children when he knew them. His uncle Alfred Kohn, who was one of the three brothers who founded the postcard company Brüder Kohn, was transported with his family, his wife Hilda, and children Georg and Lilly, who were six years and three years younger than Walter, respectively, from Vienna on 28 October 1941 to the Litzmannstadt Ghetto in Łódż, Poland.[105] They were put to slave labour and Alfred Kohn died there on 21 August 1942.

Walter's cousin Georg was forced into slave labour as a metal worker at the young age of 14 (see Figure 31 for his Litzmannstadt identity card).[106]

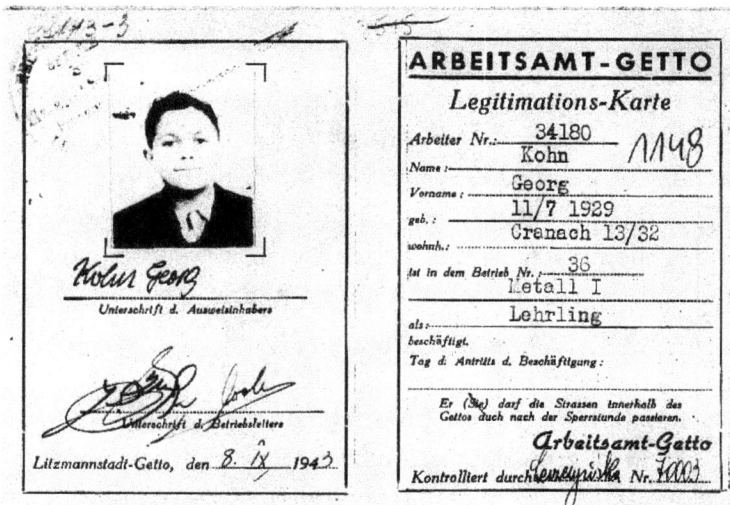

Fig. 31. The identity card of Walter's cousin, Georg Kohn, in the Litzmannstadt Ghetto, 1943. (Courtesy of the United States Holocaust Memorial Museum. Collection: Przełożony Starszeństwa Żydow w Getcie Łódzkim. Copyright: Naczelna Dyrekcja Archiwów Państwowych.)

Albert Speer, the Minister for Labour, wanted to keep the slave labour camp at Litzmannstadt running through 1944 as it was making valuable munitions and tools for the German forces. However, with the Russian armies getting closer to Łódź, Heinrich Himmler decided to liquidate the camp.[107] Accordingly, on 23 June 1944, Walter's aunt Hilda, and cousins Georg and Lilly, were transported the 80 kilometres from their residence at Cranachstrasse 13 to the Chełmno death camp which had been especially reopened to take those who had been forced to work in the Litzmannstadt ghetto. There they were murdered.[108,109] Over the next three weeks over 7,000 others were sent to their deaths in the same way from Łódź.[107]

Walter's teacher at the Chajes Gymnasium in Vienna, Emil Nohel, was also sent to Theresienstadt on 16 January 1943 from Mladá Boleslav in Bohemia. He was transported to Auschwitz on 16 October 1944 where he was murdered.[110] This was just 12 days before Walter's parents were transported there. In addition, Walter's other teacher at the Chajes Gymnasium, Viktor Sabbath, perished in the Holocaust.[11] Two of Walter's teachers at the Akademisches Gymnasium who were Jewish were also sent to Theresienstadt. David Oppenheim, who taught Latin and Greek, died there in 1943 while John Edelmann, who taught history and geography, was sent to Auschwitz on 29 September 1944 where he also perished.[110]

These tragic events gave Walter very deep feelings on persecution and genocide, on which he took the platform of the Nobel Prize to speak on many occasions. He did not classify himself as a "Holocaust Survivor" compared to the people who had to stay in the terror of the Nazi-occupied regions during the war.[6] He was, however, especially bitter that the Austrian authorities had been involved in transporting several members of his family, including his parents and cousins, to concentration camps and ghettos during the war. When his Nobel Prize was announced in 1998, and he was congratulated by the President of Austria, Walter let his views be known on this matter in no uncertain terms. His feelings on the Holocaust were even communicated to the Pope in the Vatican.

After the war, information started to filter through to Walter about the schoolfriends he had known at the Chajes Gymnasium in Vienna some six years previously (see Figure 12). Among the relatively small number of people who survived Theresienstadt was Walter's close schoolfriend Herbert Neuhaus (see Figure 32). His father Heinrich Neuhaus had been a doctor in

Fig. 32.   Herbert Neuhaus in 1938 aged 15. (Courtesy of Peter Neuhaus.)

the Rothschild Hospital and this enabled them to stay in Vienna for a longer period than many Jewish families.[111] However, in October 1943 Heinrich, and Herbert now 20 years old and working with his father, were arrested. They had been found harbouring "submarine" Jewish people who had come from Poland and were trying to enter Hungary, which was then not under the direct control of the Nazis. Herbert recounted in detail the deportation of his family to Theresienstadt:

> One month after our imprisonment, on Thursday, 11 November 1943 at 8 am, we were taken from our cell and led to a fairly big assembly room where at last we saw my mother again. After a tearful reunion we were driven in a police car to the detention center at Malzgasse 7, across the street from the hospital, where we were told that we had been assigned to a transport to the K.Z. Theresienstadt, which was to leave in the afternoon. As Vienna by that time was practically "Judenrein" (free of Jews) there were no longer any mass transports, as in the past but only small groups were deported, whenever a worthwhile number of victims had accumulated. Thus, our group consisted of 91 persons and we were placed in regular third class railroad cars and given lunchboxes by the Ältestenrat Council.[111]

The train, labelled Transport 461, left the Vienna Northern Railway Station and went via Brno and Prague. Herbert said that on arrival in Theresienstadt his family were searched and marched through the streets in the dark to their quarters.[112] Unlike Salomon and Gittel Kohn, the Neuhaus family of Heinrich, Wanda and Herbert were not deported from Theresienstadt and managed to survive the war. It is possible that, as Herbert's father was a doctor, the camp authorities found him useful (this was the case with some other doctors in the camp).[20]

The Neuhaus family returned to Vienna after the war where Herbert studied medicine. He then went to the USA where he had a distinguished medical career. He kept in regular contact with Walter and became an Assistant Professor of Medicine at the University of Illinois Hospital in Chicago and published on the lung functions of tuberculosis patients.[25]

Walter's schoolfriend Paul Sondhoff also had a remarkable escape.[113] He stayed in Austria, and volunteered in 1940 for a work camp in Traunkirchen to give some protection to his family. However, his sister Alma and mother were eventually deported on 27 May 1942 to the notorious death camp in Maly Trostinec, near Minsk, where they did not survive. Paul was then hidden in Vienna by his piano teacher and her daughter Anna Mathä:

Prof. Dr. Anna Mathä initially hid several people, including Paul Sondhoff … in a small room whereby the shutters had to remain closed and the young man had to remain in absolute immobility and dead silence. When the room was needed by the actual tenants, who had not been informed that a submarine had found accommodation there during their absence, Prof. Mathä adapted an attic room belonging to the apartment, which was accessible via a staircase. Mattresses were placed on the floor as a bed and boxes were stacked in front of them as a kind of wall for protection. Water jug, lavoir and bucket were ready on the stairs. Paul Sondhoff was only able to leave his attic when Prof. Mathä gave a signal that none of the other residents of the apartment were present.

> There were critical moments several times, for example when the water jug fell over and water began to trickle through the roof beams... A raid was made on conscientious objectors and Paul Sondhoff was in extreme danger. In need, she hid him in a deep open fireplace, where he met other hidden people.[114]

Paul Sondhoff kept a threadbare clockwork bear to remind him of life before the Anschluss and this became a prominent symbol in a major exhibition of the Holocaust held in London in 2000.[115] After the war he lived with an aunt and trained at the Technical University of Vienna as an engineer. Like several of Walter's class he went to the USA, arriving in 1951. He then worked on secret engineering projects for the US Government in New Jersey and often wrote to Walter (see Chapter Seven).

Walter's classmate Gertrude Ehrlich recalls talking to him on their very last day after leaving the Chajes Gymnasium in Vienna in June 1939:

> Someone called my name and caught up with me: it was Walt(h)er (when did you change the spelling?) Kohn, a friendly young-looking kid I'd never spoken to before. We walked together and chatted ... discussing our respective emigration plans. By the time we said goodbye I felt much better — perhaps there would be a future after all![116]

And what a future there would eventually be for both of them. Just one month later Gertrude Ehrlich escaped to the USA in July 1939 with members of her family on a boat from Holland. She took up US citizenship and studied at the University of North Carolina under the eminent number theorist and refugee from Berlin, Alfred Brauer, with whom she published.[117] He was the brother of Richard Brauer who had been one of the teachers of Walter in Toronto. She became a Professor of Mathematics at the University of Maryland and wrote important papers and books on abstract algebra.

Gertrude Ehrlich also wrote on being taught biology at the Chajes Gymnasium by Professor Elise Deiner who had published research papers in this subject.[116] Deiner was sent to Theresienstadt on 8 October 1942. There she gave a biology lecture on "The colours of creativity" on 16 July 1944 attended by over 400 inmates.[118] She was then sent to Auschwitz, on 9 October, just 19 days before Salomon and Gittel Kohn.[119]

Rudi Ehrlich (no relation to Gertrude) was the star mathematician among the several stars in Walter's class at the Chajes Gymnasium. He stayed in Vienna for the first four years of the war. In 1943 he was arrested and was due to be sent to a concentration camp when, at the last moment, he was freed through the efforts of the Italian Consulate — he had an estranged Italian father. He took up his father's surname Permutti and managed to emigrate to Trieste in Italy.[120] However, in September 1943, following the fall of Mussolini's government, the German troops occupied Trieste and he had to escape. He lost a leg in a train accident but was able to continue studying under a false name at a school in Rome, which was then occupied by the Nazis. Following the liberation of Rome in June 1944 he was able to study mathematics and eventually received a doctorate under the supervision of Francesco Severi at La Sapienza University. After further research in Naples and Bari, Rudi Permutti moved back to Trieste where he became a full professor of algebra. He made many important contributions to mathematics including Galois theory and Möbius planes.[120]

Another budding mathematician in Walter's remarkable class was Karl Greger. He escaped from Vienna to Sweden in 1938. He studied in Lund and, after the war, became an assistant at the Mathematical Institute from 1948 to 1952. He published several papers and books on number theory and probability. He wrote extensively on mathematics education and became Chair of the Mathematics Department of the High School of Education in Gothenburg.[120]

In the class one year below Walter at the Chajes Gymnasium was Otto Hutter.[121] Like Walter, he also managed to leave Vienna on the Kindertransport. A chance encounter with a friend in December 1938 directed him to the Gestapo office at the hotel Metropole in Vienna where permits to emigrate to Britain were then being granted even if no sponsor was available. After arriving at Harwich on a boat from the Hook of Holland, he first stayed at the freezing cold Butlin's Dovercourt holiday camp in mid-winter. He then lived with the Blaxill family at nearby Colchester, Essex and eventually went to university and undertook research in physiology at University College London. He made the first recordings of the pacemaker potential in heart muscle using microelectrodes. Hutter became Regius Professor of Physiology at the University of Glasgow.

There were 37 members of Otto Hutter's class at the Chajes Gymnasium and he was able to track down what happened to nearly all of them after the war.[122] Several managed also to take the Kindertransport to England and some were able to emigrate to the USA. However, at least two were murdered in the Holocaust. One of these, Kurt Mezei, survived by hiding in a cellar in Vienna until 11 April 1945, when the Russian forces were already on the edge of the city. He was then denounced and shot by a member of the SS in the last month of the war.

The Chajes Gymnasium itself was closed in 1941 and was used by the Nazi authorities to assemble Viennese Jews prior to deportation. In 1984 it was re-established in the Castellezgasse.

Chapter Six

Harvard

Walter arrived at the Jefferson Physical Laboratory, Harvard University in September 1946 (see Figure 33). The new graduate students there were quite a mix. Some of the more senior students had already spent the war at laboratories such as Los Alamos alongside many famous physicists. Others came fresh from universities all over the USA and a few, like Walter, were refugees from Europe. Toronto was no scientific backwater but it did not compare with Harvard. Walter had already published three research papers so coming to Harvard as a graduate student at the age of 23 should

Fig. 33.   Jefferson Physical Laboratory, Harvard University.

not have been too daunting for him. Nevertheless, he stated that he initially "felt very insecure and set as my goal survival for at least one year."[11]

Theoretical physics had changed a great deal over the preceding 13 years. Before Hitler came to power in 1933, Germany led the field and American expertise was sparse. However, this all changed with the exodus of so many top scientists and promising students from Germany. Several of these refugees were then gathered together in the laboratories in the USA that contributed so significantly to the war effort. Accordingly, after the war, the USA had become the world leader in the field of theoretical physics, the fame and reputation of which amongst the general public had risen enormously. Therefore, a fair number of the most brilliant young people in the USA were choosing physics as the subject they wished to follow and the competition for university places was high.

Harvard was the oldest university in the USA, being established in 1636. It had a large endowment and, in the view of many, had maintained its reputation as the leading North American university. Not only was this reputation due to the quality of the professors who were teaching there but also because of the intellectual atmosphere among its graduate community. Nearby was MIT which had close links. In addition, there was an exceptional seminar programme giving the opportunity for students to learn about the latest physics research outside of their field from the leading experts in the world. These aspects were to prove crucial when Walter changed his research direction to solid-state physics.

Following the advice from Infeld, Walter was hoping to register for his PhD research with Julian Schwinger who already had a reputation as the most promising newly appointed theoretical physicist in the USA (see Figure 34).[11] He was aged 28 and had only joined Harvard in the previous spring as an Associate Professor. Schwinger had been a research student at Berkeley with J. Robert Oppenheimer before the war.[123] There he was considered quite eccentric and was famous for working during the night so that his concentration would not be interrupted.[124] Schwinger was at the MIT radiation laboratory during the war and occasionally would visit Los Alamos to lecture. Richard Feynman who was then also aged 25 said:

Fig. 34.   Julian Schwinger. (Courtesy of American Institute of Physics, Emilio Segrè Visual Archives, Physics Today Collection.)

It was until I went to Los Alamos that I got a chance to meet Schwinger. He had already a great reputation because he had done so much work... and I was very anxious to see what this man was like. I'd always thought he was older than I was because he had done so much more. At the time I hadn't done anything. And he came and gave us lectures. I believe they were on nuclear physics. I'm not sure exactly the subject, but it was a scene you probably all have seen once. The beauty of one of his lectures. He comes in, with his head a little bit to one side. He comes in like a bull into a ring and puts his notebook down and then begins. And the beautiful, organized way of putting one idea after the other. Everything very clear from the beginning to the end... I was supposed to be a good lecturer according to some people, but this was really a masterpiece... So I was very impressed, and the times I got then to talk to him, I learned more.[125]

As is still the case in US academic programs for research in physics, Kohn had to take advanced classes right across the different areas of the subject. On Schwinger's courses Walter recalled:

> Attending one of his formal lectures was comparable to hearing a new major concert by a very great composer flawlessly performed by the composer himself. For example, his historic graduate courses on nuclear physics and waveguides given in the late 1940s consisted largely of exciting original material. Furthermore both old and new material were treated from fresh points of view and organized in magnificent overall structures. The delivery was magisterial, even, carefully worded, irresistible like a mighty river. He commanded the attention of his audience entirely by the content and form of his material, and by his personal mastery of it, without a touch of dramatization... Crowds of students and more senior people from both Harvard and MIT attended and, knowing his nocturnal working habits, I found the price of having to wait 10, 20, 30 minutes for his arrival quite trivial in comparison with what he gave us. I felt privileged — and not a little daunted — to witness physics being made by one of its greatest masters. Each of those two courses had a tremendous influence on the shape of their respective fields for decades to come, as did other later Schwinger courses such as quantum mechanics and field theory.[124]

It is often the case that a special bond is formed between a newly appointed professor and their first graduate students. Schwinger accepted an extraordinary 11 students in his first cohort. In addition to Walter, this included Roy Glauber (who would win the 2005 Nobel Prize for Physics for his research on quantum optics), Bernard Lippmann (who became well-known for the Lippmann-Schwinger integral equations for scattering processes), Kenneth Case (who already had established a reputation for his work on bomb design at Los Alamos), Bryce DeWitt (who advanced general relativity) and Fritz Rohrlich (a refugee from Vienna, like Walter, who had first emigrated to Jerusalem in 1939).[123]

Students who were soon to join the Schwinger group and would also win the Nobel Prize included the Danish-American Ben Mottelson, who would become a close friend of Walter, and Sheldon Glashow.[123] In addition, there were students from other research groups who would become friends or collaborators of Walter. This included Philip (Phil) Anderson who was working with John Van Vleck on the theory of microwave spectral line shapes (see Figures 35 and 36).[126] The wartime research on radar had provided new equipment which was enabling experimental researchers to measure the details of infrared and microwave spectra of molecules. Not only were the peaks in the spectral lines of interest but also the widths of the lines which were temperature dependent. These quantities depended on the intermolecular forces and collision dynamics of the interacting molecules. Therefore, this was a rare experimental method that had the potential to

Fig. 35. Philip Anderson, 1977. (Courtesy of American Institute of Physics, Emilio Segrè Visual Archives, Physics Today Collection.)

Fig. 36.   John Van Vleck. (Courtesy of American Institute of Physics, Emilio Segrè Visual Archives, Physics Today Collection.)

extract the details of intermolecular forces if an appropriate theory could be developed. That is what Anderson provided, most comprehensively, in his PhD research. This was not a very fundamental PhD project like some of those being pursued in Schwinger's group but it had direct links with new experimental data.

Like Walter, Anderson would soon move into solid-state physics research through the influence of John Van Vleck and they would become friendly competitors. Anderson was original and always went against the flow. On Schwinger he said: "There was this tremendous excitement and gang of people around Schwinger. I wanted to go in the other direction."[127] Anthony Leggett was to comment on Anderson: "Even when he turned out to be wrong, he was influential because he got people to think in new directions."[128] This statement recalls Pauli's famous comment on a paper he had read: "It is not even wrong!"[129] Anderson would win the 1977 Nobel Prize for Physics with Nevill Mott and Van Vleck for their theoretical research in solid-state physics.

Harvard was close to MIT and there were shared programs. Joaquin "Quin" Luttinger had done wartime research in the radiation laboratory at MIT. His PhD thesis at MIT, which was supervised by Laslo Tisza, had the title "Dipole interactions in crystals". He came over to Harvard for the famous lectures of Schwinger in the Jefferson Laboratory.[11] Schwinger was notorious for being late for his lectures and this gave the opportunity for the students from different laboratories to chat informally. In this way, Walter and Luttinger would start a friendship that would develop eventually into a close scientific collaboration.[130]

Upon his arrival at Harvard, John Van Vleck proposed a research project to Walter on the band theory of solids. However, Walter had not yet studied this area and preferred to take the advice from Infeld to work with Schwinger.[11] Like Walter, Schwinger was enthusiastic about variational methods and he at once agreed to supervise him. At Harvard, Schwinger supervised as many as 73 doctoral students, a number probably unsurpassed amongst US theoretical physicists. PhD supervisors can have different supervision techniques. Some like to check on progress with their students nearly every day. Schwinger, however, was quite the opposite. In Walter's words:

> Arranging to meet with him was devilishly hard, but when it happened — a few times a year — I found him most generous with his time and brilliant in his judgments and suggestions. It was during these meetings, sometimes more than two hours long, that I learned the most from him. He had a large old-fashioned office in the old Jefferson building.[124]

Schwinger suggested a difficult project to Walter on the quantum theory of low-energy three-body collisions of neutrons with deuterons, while freely admitting he had attempted a solution to this problem but had failed. Walter did not have much fortune on this problem either but then realised that he could directly apply a variational method in a general way to scattering processes, include those involving exchange of particles. This he would write up as a first-author paper in the *Physical Review*.[131] In his paper, which was received on 19 August 1948, he gave an acknowledgment to Schwinger "for giving invaluable counsel at all stages of the investigation".

The work would form the basis for his PhD thesis which he submitted in the very short time of two years. This paper would become the most highly cited of Walter's early publications and to date has received over 500 citations. Like his DFT, the "Kohn Variational Principle" for scattering is general and has seen many applications to molecular problems. Some of the most accurate quantum mechanical calculations on the chemical reactions between atoms and diatomic molecules have been performed with modifications of his technique.[132,133] But Schwinger's deep influence was always there as Walter explained:

> We carried away the self-admonition to try and measure up to his high standards; to dig for the essential; to pay attention to the experimental facts; to try and say something operationally meaningful even if — as usual — one cannot calculate everything *a priori*; not to be satisfied until ideas have been embedded in a coherent, logical and aesthetically satisfying structure.[123]

Not only do these statements apply to Walter's Variational Scattering Theory, they would also very much apply to his DFT. Walter subsequently wrote about the excitement of being in Schwinger's research group at Harvard:

> In one corner, at a desk, sat Harold Levine calculating away on intricate classical wave problems, totally oblivious to what was going on around him. Drifting in and out were other students anxious to catch Julian. Frequently, Herman Feshbach came over from MIT to talk about nuclear forces. A few times Freeman Dyson and Richard Feynman dropped in to talk about quantum electrodynamics. Once a letter or preprint from Tomonaga arrived and Julian said he was nervous to open it, so often had Tomonaga's thinking been almost the same as his. What great fortune for us to be there at such a time![124]

Schwinger was impressed with Walter's graduate work and offered him a postdoctoral position for three years in 1948. This also provided the option of doing some teaching which Walter readily took up.[11] The significant rise in his salary enabled him to bring Lois over to Boston where they got married on 12 March 1948 and lived at 31 Queensberry Street. Quite soon after this, their first daughter Judith Marilyn (normally called Marilyn) was born on 21 August 1949, and Walter became a very busy man.

Sidney Borowitz had a similar appointment to Walter and they shared an office. A New Yorker who had studied at New York University (where he would eventually become Chancellor), Borowitz would become a lifelong friend of Walter. At this time, Schwinger was developing his celebrated quantum theory of electrodynamics which would win him the 1965 Nobel Prize for Physics together with Feynman and the Japanese theoretician Shin'ichiro Tomonaga. Walter stated:

> In view of Schwinger's deep physical insights and celebrated mathematical power, I soon felt almost completely useless. Borowitz and I did make some very minor contributions, while the greats, especially Schwinger and Feynman, seemed to be on their way to unplumbed, perhaps ultimate depths.[11]

Walter is referring to a paper he published with Borowitz in the *Physical Review* on the electromagnetic properties of nucleons.[134] They stated: "The calculated values of the magnetic moments of the neutron and proton are in agreement with results previously obtained by Case, Luttinger, and Slotnick and Heitler using different calculational procedures." This illustrates the close link already being developed with Luttinger.

During this period, Walter also had an intriguing interaction with the theoretical chemist E. Bright Wilson at Harvard. Walter developed a variation iteration method for calculating the coefficients in wavefunctions and published this in the *Journal of Chemical Physics*. He gave acknowledgement to Wilson and his research group, and stated that his method was being applied to the calculation of molecular vibration spectra. This would be a very rare research paper published by Walter in a journal that had "chemical" in the title before he was awarded the Nobel Prize for Chemistry.[135] Wilson also would have some ideas connected to DFT before Walter's major papers on the subject.

In Toronto, Walter had enjoyed his summer research projects which got him out of the university to work on more applied topics. Cambridge in Massachusetts was already starting to be a hub for new companies spinning out from the universities and institutions in the area, and Polaroid was a good example. Cosmic ray research had demonstrated the possibilities of charged particles colliding with photographic plates and Polaroid was exploiting this process in the production of photographic images. In his summer project, Walter developed the theory on this topic and he had the opportunity to discuss his research with Van Vleck. Walter commented:

> It seems that these meetings gave him the erroneous impression that I knew something about the subject. For one day he explained to me that he was about to take a leave of absence and, "since you are familiar with solid-state physics", he asked me if I could teach a course on this subject, which he had planned to offer. This time, frustrated with my work on quantum field theory, I agreed. I had a family, jobs were scarce, and I thought that broadening my competence into a new, more practical, area might give me more opportunities. So, relying largely on the excellent, relatively recent monograph by F. Seitz "Modern Theory Of Solids", I taught one of the first broad courses on solid-state physics in the United States.[11,136]

In 1948, Walter's close friend Josef Eisinger had graduated in physics from the University of Toronto with a Master's degree. He had done research on the Raman spectroscopy of the methane molecule. He found that increasing the pressure of the gas perturbed the spectral lines of the molecule — an effect linked to Anderson's thesis research at Harvard. Josef then followed Walter to Massachusetts. He did a PhD on atomic beam research at MIT in the group of Jerrold Zacharias and even took Schwinger's famous course at Harvard.[46]

Fig. 37. Nicolaas Bloembergen, 1949. (Courtesy of American Institute of Physics, Emilio Segrè Visual Archives, Physics Today Collection.)

Walter's class on solid-state physics included some of the graduate students who were already his friends including Nicolaas Bloembergen (see Figure 37), Charles Slichter and George Pake.[11] These three were an extraordinary trio. The Dutchman Bloembergen would win the 1981 Nobel Prize in Physics for his research on non-linear optics, Slichter would be awarded the National Medal of Science for his work on Nuclear Magnetic Resonance and Pake would direct the Xerox Palo Alto Research Centre where many of the key innovations in computer technology were to be made.

This also gave Walter the opportunity to write his first papers on solid-state physics. He became interested in metallic lithium and used a wavefunction approximation to calculate its cohesive energy.[137] In another study with Bloembergen, he published on the nuclear magnetic resonance shift in metallic lithium. They made the bold conclusion "that the Wigner-Seitz-Bardeen method is not suitable for a description of the electrons on the Fermi surface of metallic lithium."[138] With Eugene Wigner and John Bardeen both to become Nobel Prize winners in Physics (Bardeen twice), this was a bold conclusion from two young physicists still building their reputations (but both to also become Nobel Laureates in due course). However, their erratum to the paper noted a mistake in their calculations and the published correction gave a better agreement with experiment.[138]

The original conclusion was, perhaps, not the greatest of starts in solid-state theory but this clearly was a hot topic.

In the summer of 1950, Walter gave a broad undergraduate course at Harvard on light and optics, electricity and magnetism, and atomic and nuclear physics. Being quite advanced for the time, a very positive student assessment on the course was written:

> Dr Kohn's lectures were clear and concise. Demonstrations accompanied every lecture and these were always received with great enthusiasm on the part of the professors and the students ... The professor often gave proofs which differed from those in the [set] text, for which we were held responsible. However, it was obvious from the beginning that his interest was not in our mastery of the mathematical problems, but in our understanding of the theory, and, in so far as possible in such a short term, to understand how physical concepts evolved. It was not infrequent that Dr Kohn read from some source material giving such direct quotes as Newton's, relating his discovery of the refraction of light... The content of lectures was at all times stimulating.[139]

Walter had now published several papers, had done some teaching which had been positively received, and had networked well. He also had a young family and the time was right for him to find his first independent appointment. However, the large number of physicists coming out of the war-related laboratories had already taken up most of the available positions at universities. In addition, the competition was very strong with many PhDs in theoretical physics on the job market, not just from Harvard and MIT, but from Columbia, Cornell, Chicago, Princeton, Stanford, Berkeley, Caltech and elsewhere. Walter applied for several posts in US universities but was not successful initially. He had been a Canadian citizen since 1943 and, with his wife Lois also Canadian, he applied to some Canadian universities but had no luck there either.[11]

By 1950, Walter had been away from Europe for over ten years. In that time so much had happened. His parents, other members of his family and some of his school teachers had perished in the Holocaust. His sister Minna had got married and he had not met his new brother-in-law Franz Pixner. She had also started to revive the artistic postcard business Brüder Kohn of her father. Through a remarkable journey, he had received an education that was putting him at the forefront of modern physics. He felt that a sabbatical in Europe would be exciting and applied to the US National Research Council to work with the Nobel Laureate Wolfgang Pauli who had recently returned from Princeton to the ETH in Switzerland.[54,129]

Then, as often happened in his life and career, Walter had a stroke of luck which he recalled:

At this point a promising possibility appeared for a position in a new Westinghouse nuclear reactor laboratory outside of Pittsburgh. But during a visit it turned out that US citizenship was required and so this possibility too vanished. At that moment I was unbelievably lucky. While in Pittsburgh, I stayed with my Canadian friend Alfred Schild, who taught in the mathematics department at the Carnegie Institute of Technology (now Carnegie Mellon University). He remarked that F. Seitz and several of his colleagues had just left the physics department and moved to Illinois, so that he thought there might be an opening for me there. It turned out that the Department Chair, Ed Creutz, was looking rather desperately for somebody who could teach a course in solid-state physics and also keep an eye on the graduate students who had lost their "doctor-fathers". Within 48 hours I had a telegram offering me a job![11]

Chapter Seven

Pittsburgh

The Carnegie Institute of Technology was founded by Andrew Carnegie in 1900 through a major donation to establish a technical institute in Pittsburgh. Carnegie had emigrated from Scotland and had made a fortune through steel making in Pittsburgh towards the end of the 19th century. Carnegie's initial vision was to create an institution where men and women from Pittsburgh would learn practical crafts and trades. The original name was Carnegie Technical Schools but with the demand for more advanced courses emerging, including bachelor degrees, the institution was then named the Carnegie Institute of Technology. Fine buildings were constructed and, in the first half of the 20th century, the Institute began to establish a reputation for research especially in physics, chemistry, metallurgy and engineering.[140] This progress was accelerated after the Second World War with significant funding becoming available following on from the huge contributions of science and technology to the war effort. In the USA, several other major technology institutes had also emerged, with Caltech and MIT being the most famous.

Carnegie Tech did not have the Californian glamour of Caltech, nor the deep academic esteem of MIT in Cambridge, Massachusetts, but its reputation was growing due to the acquisition of some top-class scientists. The Physics Department was a good example, where Fred Seitz had become the Chair. As a PhD student with Eugene Wigner at Princeton, he had developed a quantum theory of crystals and had written the leading textbook on solid-state physics which Walter had used in his class at Harvard.[136] Seitz did important metallurgy work in the war period in Wigner's group which was producing nuclear reactors.[141] Also present in the Physics Department at Carnegie Tech was Otto Stern who, with Gerlach, had done a famous experiment demonstrating the quantisation of electron spin. After leaving Germany in 1933, he had won the Nobel Prize for Physics in 1943.

Fig. 38. Ed Creutz, 1970. (Courtesy of American Institute of Physics, Emilio Segrè Visual Archives, Physics Today Collection.)

In 1949, Seitz accepted the headship of the physics department at the University of Illinois. There he would build up a department with an outstanding reputation for solid-state physics.[141] This included John Bardeen, and his students Leon Cooper and Robert Schrieffer, who would jointly win the Nobel Prize for their theory of superconductivity.

A major appointment by Seitz to Carnegie Tech was Ed Creutz (see Figure 38). He had worked with Eugene Wigner at Wisconsin and then at Princeton. In an "Anticipatory Obituary" written in 1996, Creutz wrote:

On my third day at Princeton I was invited to give a short report on my thesis. There were usually two or three speakers at these "Journal Club" meetings. This time the speakers were Niels Bohr, Albert Einstein, and Ed Creutz! To be on the same program with these two giants of scientific accomplishment was breathtaking. Just before the meeting began, my sponsor,

[Lewis] Delsasso, asked me, "Say, Creutz, have you met Einstein yet?" I had not. Delsasso took me over to where Einstein was sitting in sweatshirt and tennis shoes, and said, "Professor Einstein, this is Creutz who has come to work on our cyclotron." The great man held out his hand, which seemed as big as a dinner plate, and said in an accented voice, "I'm glad to meet you, Dr. Creutz." I managed to wheeze out, "I'm glad to meet you, too, Dr. Einstein."[142]

It was at this very event in February 1939 that Niels Bohr brought to the USA the first news of Otto Hahn's discovery of nuclear fission that would lead, after a key letter from Einstein to President Roosevelt, to the Manhattan Project. Creutz subsequently played a major role at Los Alamos in the design of the lens to initiate the explosion of nuclear weapons.[142] In appointing Creutz to Carnegie Tech after the end of the war, Seitz was anticipating that he would be able to attract significant funding for nuclear research. This was the case and Creutz and colleagues built one of the largest cyclotrons. Creutz replaced Seitz as Chair of the Department of Physics in 1949 and Walter was one of his first appointments to an Assistant Professorship. Creutz was to become a key mentor in the career of Walter Kohn. Pittsburgh also had the advantage for Walter and Lois in that it was one of the closest major US cities to Toronto where Lois had family and Walter many friends.

Creutz wrote to Walter, who was still at Harvard, on 14 July 1950 with a welcome suggestion:

One of our Navy contracts, under which work on the mobility of electrons in diamond and Hall Effect in diamond has been studied, is terminating and I should like to renew it. It occurred to me that you might wish to do some work or have some students working under this contract, and I should therefore like to get your ideas on it before I send in the proposal to the Navy. At present we have a student measuring the luminescent efficiency of single crystals when bombarded by electrons, X-rays and ultra-violet. This is one project I wish to put under the new proposal. Also a continuation of the measurement of the mobility of electrons in diamond by measuring the rise time of pulses produced by gamma rays will be continued. If you have other suggestions that might be included in the project, please let me know, if possible within the next week.[143]

So, at once, Walter was being offered easy funding for research in solid-state physics arising from the excellent contacts of Creutz with the US defence establishment. In the 1950s, funding for physics research in the USA was not a problem. Carnegie Tech, and its successor Carnegie Mellon University, were to play important direct and indirect roles leading to Walter's Nobel Prize (see Figure 39).

A few weeks after accepting the offer at Carnegie Tech, Walter heard that his application to the National Research Council to fund a one-year

Fig. 39. Carnegie Institute of Technology, ~1950. The Department of Physics is on the left of the main quad. (Creative Commons Attribution-Share Alike 2.0 Generic licence.)

Fig. 40. Niels Bohr. (Courtesy of American Institute of Physics, Emilio Segrè Visual Archives, Margrethe Bohr Collection.)

fellowship with Pauli at the ETH in Switzerland had been approved. However, as he explained in a letter to Niels Bohr, a visit by Aage Bohr to Harvard had convinced Walter that his year abroad would be better spent in Copenhagen at the Institute for Theoretical Physics directed by Aage's famous father (see Figure 40).[144] Niels Bohr then wrote a supporting letter to the National Research Council stating: "Dr Kohn shall be most welcome to work with our group and we shall do our best to make his stay among us a profitable time for his further studies."[145] Nobody would refuse a request from Niels Bohr and the funding for a year in Copenhagen was approved. Walter was to be paid a stipend of $3,750 plus a travel allowance of $500.[146] As the Fellowship was involved with training, no tax was to be deducted. At the time, this was a generous offer.

Walter also negotiated with Creutz, who knew Bohr well, to be granted a one-year sabbatical which would begin only one semester after his starting at Carnegie Tech.[11] The condition was that Walter taught a compressed course in solid-state physics in the fall of 1950 before going on to Copenhagen. Research in Bohr's Institute for Theoretical Physics (see Figure 41) was then mainly on nuclear physics with little interest in solid-state physics. Walter anticipated that his work on nuclear scattering would therefore have to be

Fig. 41. Institute for Theoretical Physics, University of Copenhagen, October 1951. Niels Bohr is 5th from the left in the first row. In the second row are Walter Kohn, Ben Mottelson, Aage Bohr and Vachaspati, 7–10th from the left. (Courtesy of the Niels Bohr Archives.)

the emphasis, at least for his year in Copenhagen. He was very keen to return again to Europe, for the first time after the war, and he was looking forward to meeting up again with what was left of his family.

Walter, Lois and baby Marilyn arrived in Copenhagen right at the end of 1950. Lois particularly enjoyed living in Copenhagen. In her subsequent obituary it was stated:

> The young family had two magical years in Copenhagen. Lois embraced the Danish culture; she loved the Danes' warmth and charm, their artisanal work, their staunch loyalties, and not least their bravery in saving their Jewish citizens in World War II.[92]

Through discussions with a visitor in Copenhagen from India, Vachaspati, Walter became interested in superconductivity, despite the minimal interest in solid-state physics at the Institute for Theoretical Physics. Vachaspati was the first of several Indian theoretical physicists who would collaborate with Walter. They wrote a short paper for the *Physical Review* with the title "A difficulty in Frohlich's theory of superconductivity" and indicated that Frohlich's use of second-order perturbation theory could lead to numerical difficulties for excited oscillator states.[147] However, John Bardeen also then wrote a paper on superconductivity which criticised the conclusions of Walter and Vachaspati.[148] Walter wrote back to Bardeen about this and he responded to dispute Walter's conclusions.[149] Walter wrote back still disagreeing with Bardeen.[150] If he thought he was right in science, Walter would say so even to his most eminent competitors. Bardeen was quite soon to produce a new theory of superconductivity with his research students Cooper and Schrieffer (the Bardeen-Cooper-Schrieffer (BCS) theory) that would win him a second Nobel Prize and this work was perhaps influenced by his discussions with Walter.[151]

Walter and Lois enjoyed Copenhagen so much that he asked the National Research Council to extend his Fellowship by a year. However, they did not have the funds to allow for this.[152] He was then awarded the prestigious Oersted Fellowship from Bohr's Institute to extend his Fellowship into 1952.[11] Accordingly, in the fall of 1951, Walter wrote to Creutz to request another year of leave. Creutz replied saying this was approved, but only after a lot of work on his part and he was not sure who would give Walter's course of solid-state physics.[153] Walter responded in very warm terms: "I can hardly find adequate words to tell you how much I appreciate what you have done for me again. I only hope that I shall be able to make use of the additional time in a way worthy of the confidence you have shown in me."[154] Walter also mentioned that Copenhagen was hoping a new "European Laboratory" would be established there and this perhaps implied he was thinking of a move in that direction. Although Walter was not being paid by Carnegie Tech when he was in Copenhagen, Creutz and his colleagues must have been thinking that they might not see their young Assistant Professor there again. However, the European Laboratory went to Geneva where it was most famously called CERN.

Walter's enthusiasm for Copenhagen was extending to the former students he had met in Harvard. Frederic de Hoffmann was one year younger than Walter and was also born in Vienna.[155] He was brought up in Prague and managed to escape to the USA at the very late date of 1941. De Hoffmann then worked at Los Alamos during the war and returned there after his Harvard graduate research. He joined the group of Edward Teller and played an important role in the development of the hydrogen bomb. In the subsequent official history of the Atomic Energy Commission, it was stated that "Teller had [in Frederic de Hoffmann] an able and shrewd scientific aide of high managerial and political ability."[155] De Hoffmann wrote to Walter asking many questions about the possibility of a visiting position in Copenhagen.[156] Walter replied: "The members have attractive offices, there is a very warm spirit in the place and, living a few thousand miles from the boiling point of US Physics, the atmosphere is pleasantly relaxed."[157] De Hoffmann was to reappear several times in Walter's career. He was to play a key role in expanding the University of San Diego and became President of the Salk Institute in La Jolla. In 1989, de Hoffmann had a bypass operation with infected blood and he tragically died of AIDS at the age of 65.[155]

Travel in Europe was easy, and Walter was able to visit Rudolf Peierls in Birmingham in 1951 and Harrie Massey in London. Walter had nearly done his graduate work with Peierls and had a regular correspondence with him throughout his career. On this trip, Walter left his briefcase with many precious notes on a train and was very surprised and delighted when it was returned.[158]

In Copenhagen Walter also wrote a paper on the validity of the Born approximation.[159] In 1926, just after the publication of the Schrödinger equation, Max Born in Göttingen showed how to apply the theory of wave mechanics to atomic scattering processes in an approximate way. Part of his theory involved using the probability density (the absolute square of the wavefunction) to calculate the scattering amplitude. It was his idea of probability density, hotly disputed by both Einstein and Schrödinger, which was to win Born the Nobel Prize for Physics in 1954.[72] Walter rewrote Born's derivation in terms of a series and established bounds on its accuracy. In more recent years, like Walter's variational scattering theory, modifications of the Born approximation have been used in calculations on the quantum dynamics of chemical reactions.[160]

Being based in Copenhagen, Walter had the opportunity to travel to other nearby European countries in 1951. Bruno and Hertha Mendel had been so supportive to Walter when he lived in their house in Toronto after being released from internment. Bruno Mendel had moved to the University of Amsterdam in 1950 to the Chair of Pharmacology and Walter was able to visit him and Hertha there. He was also able to go back to Vienna for the first time since his departure on the Kindertransport in August 1939.[161] In this war-ravaged city, which had severe food shortages and was partly under the control of Russian forces, he met his sister Minna for the first time in 11 years and her husband Franz.

In the summer of 1951 Walter was invited to lecture at the first summer school at Les Houches near Chamonix in France.[11] Since this first school, Les Houches has become a major scientific centre not just for summer schools but also for winter schools in the snow (as the author of this book can confirm). Walter was a last-minute substitute and he gave some lectures on solid-state physics. This was quite daunting as Wolfgang Pauli was in the audience. Also present was Res Jost, who was an assistant to Pauli at the ETH in Switzerland (see Figure 42). Jost had already done fundamental theoretical work on quantum scattering theory which continued to be an interest for Walter. Jost had found a new type of solution to Schrödinger's equation when applied to scattering processes and this became known as the Jost Function.[162]

Fig. 42. Res Jost, 1983. (Photograph by Hermann Gagg. From the collection of the ETH-Bibliothek. Creative Commons Attribution-Share Alike 4.0 International.)

Jost visited Copenhagen and started a collaboration with Walter on a theory for inverting scattering data to obtain the interaction potentials between nuclei. Jost was then appointed to a prestigious Senior Fellowship at the Institute for Advanced Study in Princeton where Einstein was based. They would write up this work when Walter made a quick trip to the Institute early in 1952 at the invitation of the famous director J. Robert Oppenheimer.[163] Walter and Jost would remain close friends until the death of Jost in 1990.

Walter always had deep feelings for the University of Toronto and very much hoped to be offered a position back there. His Assistant Professor post at Carnegie Tech did not have tenure and the salary was not substantial. Lois had family close to Toronto and wanted to return home. Samuel Beatty had been so helpful to Walter when he started as an undergraduate at Toronto in 1942. Accordingly, in September 1951, Walter wrote to Beatty asking about the possibility of a post in his Department of Applied Mathematics. Beatty, however, responded in less than positive terms:

> It looks as though we shall watch the present set up [in Applied Mathematics] for a year or two in the hope that such lacks as we find can be made up for by appointing a young PhD as Lecturer... I shall keep in touch with you in case a real opening presents itself by next Spring, which is the last time at which I expect to have any responsibility. I am sorry I cannot be any more definite at present.[161]

Beatty's letter implies that he was still thinking of Walter as a "young PhD" and not the somewhat more mature Assistant Professor he now was. A very disappointed Walter replied at once in what was almost a manifesto for science:

> Since my attachment to Canada is very strong, I may perhaps be permitted to express a few personal thoughts in connection of your description of the Applied Mathematics Department.

I believe that every thinking person is agreed that research in physics is playing an ever increasing role in our society, and so it is not surprising that most countries which look ahead into the future are giving it enormous public and private support; witness the USA, Britain and also the USSR.

I have been glad to notice during the last year or so the enlightened international fellowship program put into being by the Canadian National Research Council. I am sure that this is providing an invaluable stimulus for science in Canada and will help materially to keep the country abreast of current developments. Especially for a relatively young country like Canada it is obviously of extreme importance not to fall behind and become a mere follower of other countries with more initiative.

But these activities in Ottawa and Chalk River are clearly not enough. It is up to the universities to educate young Canadians who can take up and carry on by themselves the research at the National Laboratories. It is also essential to have significant research work being conducted at the universities themselves, without which it is impossible either to attract the best teaching personnel or to provide that creative atmosphere which is so essential for the young promising student. Finally it is indispensable to promote international contacts and exchange for science which has always had a record of cutting across frontiers, if more international today than ever. Particularly since coming to Europe, I have been struck by the true internationalisation of all scientists...

During my time at Toronto, certainly in the field of theoretical physics, the University met in a splendid way the responsibilities about which I have talked above. It was therefore with a feeling of some sadness when I read your description of how today your limited staff is trying, but of course is physically unable to carry on the old tradition, where meanwhile a new outlook is urgently required. I was particularly disheartened that if necessary any shortcoming of the present system might be remediated by the appointment of a young PhD lecturer. If you had written me that the Department did not require my services because you were about to get a more capable and experienced man, I would have been sad for myself but glad for Canada and Toronto. As it is I can't help but being deeply disappointed. I might add that I shall from now on stop looking back to Canada and try and establish myself firmly in the US. I shall do this with many regrets, but at least in the knowledge that I am one of thousands of young Canadians who have had practically no other alternative.[164]

However, despite these strong misgivings, Walter still retained aspirations for Toronto and it would not be long before he was writing there again. He also had received several letters from the Canadian Department of Veteran Affairs in Ottawa. They had given him some support when he took up graduate work in the USA and were expecting him to return to Canada, it seems, as part of the "deal". It has to be said that Walter did his best to return but it did not happen.

Walter's stay at the Institute for Advanced Study at Princeton in 1952 to collaborate with Jost gave him an opportunity to make a quick visit back to Carnegie Tech and show his intentions to continue there. In May 1952 Creutz wrote to Walter to say that he wanted him to teach the Theory of Solids graduate course in the first semester of the 1952–53 academic year and also teach undergraduate Kinetic Theory and Thermodynamics in the same semester.[165] In the second semester Walter was asked to teach a graduate course in Thermodynamics and an elementary course in Mechanics. This was an extensive teaching load for the coming academic year and it was to weigh heavily on Walter's efforts to find an alternative appointment. However, as he had been away from Carnegie Tech for almost two years it was really no surprise that he was asked to teach several courses. Creutz also informed Walter that his salary was to receive a welcome increase to $4,800.

Walter's sister Minna was able to visit Walter and his family in Copenhagen in July 1952 just before they were to return to the USA.[166] She wrote in good English to say that her children were much looking forward to seeing their aunt and uncle again and she asked "Has the baby arrived yet?" Ingrid Elizabeth was born on 4 July 1952.

On 16 August 1952, Walter, Lois, Marilyn, who was nearly three, and baby Ingrid left Europe to return by boat to the USA. This was to prove a difficult few months for the young family. Lois and the children went to live with her parents in Toronto for an expected short period while Walter sorted out housing in Pittsburgh. However, Lois contacted a serious joint infection which required the prescription of the newly discovered penicillin. She was confined to her bed for several months and needed extensive treatment. This added financial pressure on the family. Walter was starting a heavy teaching programme and he had heard about some financial difficulties at Carnegie Tech. In addition, his position was still an Assistant Professorship and he did not have the security of tenure. He therefore sent letters to friends and colleagues about the possibility of finding a new appointment. In October 1952 he wrote a heartfelt letter to his close friend Sid Borowitz detailing all of his concerns.[167]

A year before, Walter had written the very critical letter to Samuel Beatty stating he would not be looking again for a position again in Canada.[164] He was, however, now softening this view. He kept in contact with Cecilia Krieger who was originally from Vienna and was the first woman to receive a PhD in mathematics in Canada. She was concerned about job losses in the mathematics department, where she worked, at the University of Toronto. The professors Infeld and Stevenson, who had been mentors for Walter when he was a student, had resigned. She encouraged Walter to seek the advice of Professor Harold Coxeter, whose mathematics lectures he had enjoyed.[168] So on 28 November 1952, Walter wrote to Coxeter, putting himself forward again for an appointment in Toronto:

Perhaps you know that ever since I graduated from the U. of T. I have been interested in eventually going back and teaching there, and that a number of times my negotiations with Dean

Beatty reached a fairly advanced stage. I understand that at present you are a member of a committee which administers the Department so I thought I would contact you informally to ask if there will be a vacancy in Applied Mathematics which I could perhaps fill.[169]

On 17 December Coxeter responded rather formally:

Dear Kohn. Several of us would be happy to see you back in Toronto. However, we do not have much hope of managing this, because the President does not see his way to expand our department for the present... Sincerely yours, H.S.M. Coxeter.[170]

Walter then applied elsewhere in Canada. Gerhard Herzberg had been appointed as Director of the National Research Council Laboratory in Ottawa. A refugee from Germany, he was a molecular spectroscopist who would win the Nobel Prize for Chemistry in 1971. In Walter's critical letter to Samuel Beatty he mentioned he had been impressed by the creation of a National Laboratory in Ottawa. Herzberg, however, responded to a letter from Walter stating that there were very few appointments for theoretical physicists in his laboratory — the main emphasis was chemistry.[171] Little did Herzberg know that he was turning down a future Nobel Laureate in Chemistry!

Walter also wrote to the University of British Columbia but this did not prove fruitful. In addition he wrote to Luttinger who was now at the University of Wisconsin.[172] Walter had known Luttinger as a graduate student at Harvard and, although there were no positions available in Wisconsin, in a strange way this allowed for Walter and Luttinger to re-engage and start a highly productive collaboration on theoretical solid-state physics.

Philip Wallace was a professor at McGill University in Montreal, Canada. He was an expert on the band structure of graphite and had been supervised for his PhD by Walter's mentor Infeld. Through this contact, in February 1953, Walter was offered a position of Professor of Mathematics at McGill with a salary of 6,000 Canadian dollars per year.[173] McGill had become well known for physics as Ernest Rutherford had done some of his pioneering early research on the structure of the atom there in the early 1900s. Walter and Lois then visited McGill and he gave the impression that he was very much inclined to accept their offer. However, in a previous letter in January 1953, he had already indicated to Wallace some concerns.[174] One aspect was there did not seem to be a strong link between mathematics and physics at McGill. In Canada, areas such as theoretical quantum mechanics were still considered to be appropriate for mathematics departments whereas in the USA this field was now central in physics. Walter also wrote of additional worries:

There are two related points we have heard about which are troubling us quite a bit. You have probably heard that I am Jewish. My wife is Protestant and we are bringing up our children in a very liberal Jewish tradition. Now we have been told that McGill has "numerus clausus" for Jewish students which in our opinion is bad enough in itself and probably indicates an intolerant attitude on the part of some powerful forces in the University. I would be very happy if you could shed some light on this point for us. In this connection, I would be interested to know if there are a substantial number of Jewish staff members at McGill.

The other point is that Montreal is said to be strongly segregated with various national and religious groups. Our friends have for the most part been University people of all denominations and nationalities and we wonder if we would find similar circles in the atmosphere of segregation which is supposed to dominate the city.[174]

Wallace wrote back at length to emphasise that in his opinion there was a very tolerant atmosphere at McGill.[175] Walter responded positively and continued the negotiations, although subsequent comments he made referred to his reservations on these points. McGill continued quotas on the number of Jewish students ("numerus clausus") up to the 1960s.

Late in 1952 Walter heard from his close friend Josef Eisinger, who always called him "Rappa" after the family name of his mother.[176] Eisinger, who now had his PhD from MIT, was offered a postdoctoral position at Rice University. However, he had trouble getting a visa as it was the height of the McCarthy era in the USA. Eisinger had made a visit by boat to Palestine to see his parents, who had escaped there from Vienna at the start of the war. His boat had stopped at several North African ports and in Egypt, and this put him under suspicion with the US authorities. Eventually his visa came through and he was subsequently offered employment at the famous AT&T Bell Laboratories (Bell Labs) in New Jersey.[46]

Walter also had his sights on Bell Labs. This esteemed laboratory had become world famous through the invention of the transistor in 1947 by Shockley, Brattain and Bardeen (see Figure 43). This work would win the

Fig. 43. (From left) Shockley, Brattain and Bardeen, inventors of the transistor, 1948. (Courtesy of American Institute of Physics, Emilio Segrè Visual Archives, Physics Today Collection.)

Nobel Prize in Physics for this trio in 1956. The Physical Research Laboratory at Bell Labs would win ten Nobels in physics for research ranging from the microwave background of the universe to lasers and charge-coupled devices. There was no place more exciting for research in the 1950s.

Walter applied to Bell Labs and was interviewed by no less a person than Shockley in December 1952, who offered him a permanent position.[177] John Bardeen was the theoretician connected with Shockley's group and he had just left for the University of Illinois in 1951. Walter had already corresponded with Bardeen over their publications on the theory of superconductivity and Shockley may have anticipated that Walter, who had references from very strong supporters in Schwinger, Bohr and Van Vleck, would be a suitable replacement for Bardeen. Nevertheless, this offer was rather remarkable considering that Walter had not yet published a major paper in solid-state physics. The annual salary offered by Bell Labs was $8,400, which was a considerable increase on what Walter was getting at Carnegie Tech. His proposed responsibilities included "theoretical physics of the fundamental mechanisms of semiconductors, as discussed with you by Dr Shockley".[177] So this was Walter's second offer of a permanent post in the period of a few months and it was very attractive.

There is nothing like more than one offer to a brilliant Assistant Professor from other institutions to get university leaders and committees into action. Following the offers to Walter from Bell Labs and McGill, Carnegie Tech rushed through a promotion for him. Creutz wrote to offer Walter an Associate Professorship at $6,000 for ten months a year for five years.[178] This was a substantially better offer (with the possibility of additional summer salary) than the one from McGill and he decided to accept it.

There was still the matter of Bell Labs. Walter skilfully negotiated with Shockley to turn his permanent offer into one in which he would have a visiting position in the summer at a stipend of $700 a month.[179] This turned out to be an important influence on Walter's subsequent career due to the outstanding research groups in solid-state physics at Bell Labs. His former Harvard friend Anderson was already on the staff, as was Gregory Wannier who had developed a set of mathematical functions for describing the solid state that Walter was to use in his own calculations. In addition, there were then others including Charles Kittel and Conyers Herring, whom Walter could potentially interact closely with.

Therefore, following his low point in the fall of 1952, by the spring of 1953 Walter had quickly seen a huge turnaround in his personal situation. His wife Lois had recovered from her illness, the family was back in Pittsburgh, his position at Carnegie Tech was secured and the prospects for the future were bright. With the possibility of a position in Canada now a receding memory, Walter and Lois made applications for the long process of US citizenship which eventually came through in 1957.

This was finally the opportunity for Walter to publish an extensive set of papers in solid-state physics. Norman Rostoker was a postdoctoral researcher at Carnegie Tech who had started an interaction with Walter in his first semester in 1950. With Walter back full-time, they developed a theory for the band structure of the energy of electrons in periodic solids.[180] The method exploited Walter's variational scattering method. Rostoker initially conducted the numerical applications of the theory on metallic lithium. Then their general method was extended to copper. The American Physical Society meetings had become a vital medium for communicating new results and Walter presented their new theory there and at a prestigious Gordon Conference on the Physics and Chemistry of Metals held in New Hampton. The Gordon Conferences, held in somewhat isolated and primitive boys' boarding schools in New Hampshire for a whole week, had become a major series for announcing new results. Jan Korringa from Leiden had also developed a related theory and the method became widely known as the KKR technique (short for Korringa-Kohn-Rostoker).

With Rostoker, Walter had found a procedure for doing research that was going to be the trademark of many of his most important papers. He would now often collaborate very intensively with another like-minded researcher to thoroughly solve a key problem in solid-state physics. He particularly liked to address problems on which there were new measurements and develop a theory to explain the experimental findings. He would find that his summer interactions at Bell Labs would be particularly fruitful in this regard.

Walter's project during his first summer at Bell Labs in 1953 was suggested by Shockley and involved calculating the damage done to silicon and germanium crystals by energetic electrons. This related to the important problem of radiation damage of semiconductors in outer space. Walter calculated that the likely displacement of nuclei from radiation interactions was smaller than had been previously expected. With Bell Labs paying his travel expenses, he presented these results at a meeting of the American Physical Society in New York in January 1954.[181] His quick completion of this project impressed Shockley and colleagues at Bell Labs, and Walter was subsequently invited to work there for each of the next 12 years. His appointment letter from Bell for the next summer of 1954 increased his stipend significantly to $850 per month.[182]

Walter was delighted to meet up again with Luttinger at Bell Labs in the summer of 1954 (see Figure 44). Luttinger had worked with Pauli at the ETH in Switzerland directly after the end of the war and had also interacted with Jost. He had carried out a quantum electrodynamics calculation of the anomalous magnetic moment of the electron at the same time that Schwinger did his calculations. Luttinger had held positions at Princeton and Wisconsin and now was a professor at the University of Michigan. Luttinger had got to know Walter at Harvard when Schwinger was lecturing there and they had kept in contact since then — Walter had even asked Luttinger about possible positions in the fall of 1952.

Walter and Luttinger at once started a major collaboration in developing an effective-mass theory for charge carriers in semiconductors such as silicon and germanium. Their theory was general. It treated optical and magnetic

Fig. 44.   Quin Luttinger. (Courtesy of American Institute of Physics, Emilio Segrè Visual Archives, Physics Today Collection.)

effects and could be applied to impurity states that are central to many electrical properties. Their paper was published in 1955 in the *Physical Review* with the title "Motion of electrons and holes in perturbed periodic fields".[183] The address for the paper was Bell Laboratories, with acknowledgements given to Wannier and Herring. There was no mention of Carnegie Tech. The Kohn-Luttinger paper has to date received over 2,700 citations. After the DFT papers with Hohenberg and Sham, this is Walter's most highly cited paper.

Walter was always deeply pleased to hear from his old school friends he had known in Vienna (see Figure 12). Paul Sondhoff, who had managed to survive the war hidden in Vienna, wrote in German in June 1953.[184] He said that he had started a new job as a tool designer in New York. He mentioned other school friends who were now in New York including Heinz Zoldester, Theo Kupfermann and Gerda Sterzer. Gerda's brother Ernest had a remarkable escape from death being interned first in Theresienstadt and then Auschwitz while needing a regular dose of insulin.[185] Nevertheless, after a brutal death march, he survived the war and was brought to the USA by his sister. Heinz Zoldester stayed in Vienna and was able to get a PhD there after the war. He was yet another member of Walter's class at the Chajes Gymnasium in 1938 who went on to become a distinguished professor in the USA, in this case in German Literature. Letters such as these brought back to Walter his school days of pre-war Vienna and, as he grew older, these memories became more important to him.

In the short period of 1954–55, Walter and Luttinger published jointly five papers in the *Physical Review* on topics that also included donor states in silicon, electron transport phenomena and cyclotron resonance, all subjects of considerable experimental interest at Bell Labs. These papers placed Kohn and Luttinger firmly amongst the leaders of theoretical solid-state physics in the 1950s. After showing great promise for several years, Walter had finally made it in this field. His hero Fred Seitz wrote asking if Walter could write a review on the effective-mass theory for his *Solid-State Physics* series.[186] Walter had always held Seitz in high esteem ever since he had used Seitz's book for his first course he gave on solid-state physics at Harvard. He readily agreed.

In 1954 Walter heard from the Physical Society of Japan who were intending to publish a set of the most important papers in general scattering theory.[187] They listed papers by Møller, Lippmann, Schwinger, Gell-Mann, Brueckner, Dyson, Salam, Jost and Kohn. So not only had Walter by now established an international reputation for himself in solid-state theory, he had done the same in scattering theory. Around that time in his Curriculum Vitae he emphasised he was working in both of these fields.

During this period, Walter also became concerned about the loss of the investments and property of his family in Europe during the war. On behalf of himself and his sister Minna, he contacted the Rotterdamsche bank in Amsterdam about some investments his great uncle Bernhard Rapaport made before the war.[188] However, the negotiations proved complicated and he concluded that the time and money spent would bring little reward.

Walter always had a very high regard for Jost. In November 1954 he wrote to him at Princeton with a concern:

> Recently I have heard, in fact from several sources, that you have made the dreadful decision (from our point of view) to return to Europe. What a mistake! … Quin and I have been engaged in something of a semiconductor orgy… As you may suspect, the results have been utterly trivial on any reasonable scale, but of course not by the standards of semiconductor physicists.[189]

Walter never lost his enthusiasm for sabbaticals and soon was to write on 9 May 1955 to Professor Busch at the ETH asking if it may be possible to arrange a Visiting Professorship there for the summer semester to enable his collaboration with Jost to continue.[190] However, after several more queries he was informed there were no funds available. Sadly, his collaboration with Jost would not be continued.

Walter was learning about the promise of using computers in his research and had correspondence with Seymour Koenig of IBM about solving a 6 × 6 determinant which came up in the Kohn-Luttinger theory. He stated: "Thank you also for your offer to do something with my 6x6 determinants. If and when I really have to have them evaluated — of course the entire object of theoretical physics is to circumvent such evaluations."[191]

However, his Institute was soon to acquire one of the first major computers for academic research and would become in due course, as Carnegie Mellon University, a leader in computational science, especially after the arrival of John Pople. Walter did not seem to be aware at the time that computers would revolutionise theoretical physics and chemistry and would help to win him his Nobel Prize.

Josef Eisinger wrote to Walter in November 1954 to say that "Sir Francis Simon" was coming to lecture at Bell Labs.[192] He was the same Francis Simon who had visited his daughters who had stayed in the Mendel's house during the war at the same time as Walter and Josef. Josef said he had not realised how important Simon was. His letters to Walter always started with "Dear Rappa" and were signed off by "Terry". Walter was to act as a professional referee for Josef and helped him get a Guggenheim Fellowship to visit Bohr's institute in Copenhagen.

In 1955, on behalf of a joint committee of the American Institute of Physics and the American Physical Society, Fred Seitz sent a questionnaire to several chemical physicists and solid-state physicists enquiring as to whether papers on solid-state physics which would previously have been submitted to the *Physical Review* should now go to the *Journal of Chemical Physics*. This latter journal had been published by the American Institute of Physics since 1933 and also covered all the areas of physical chemistry. Walter had published nearly all his papers in the *Physical Review*, and just one in the *Journal of Chemical Physics*, and did not like the proposal. He replied: "Solid-state physics has closer ties to other branches of physics than to chemistry and would be damaged if these ties were weakened."[193] No change occurred with the journals but, given Walter's comment, it is ironic that the *Journal of Chemical Physics* would publish many of the papers that would develop and apply DFT.

Walter's influence in his department at Carnegie Tech was growing. He invited Harry Jones, a Fellow of the Royal Society from Imperial College London, to visit and consider a professorial position at Carnegie Tech. Originally a student of Nevill Mott in Bristol, UK, Jones was doing interesting work on the theory of metals. However, he had just been awarded a chair at Imperial and declined the offer.[194] It seems that, after this false start, Walter was to be making regular recruitment offers to colleagues for the next 30 years.

Fig. 45.   Keith Brueckner, 1960. (Courtesy of American Institute of Physics, Emilio Segrè Visual Archives, Physics Today Collection.)

With his attempt to go to ETH for a sabbatical unsuccessful, Walter arranged a sabbatical at the nearby University of Pennsylvania in the fall of 1957. Luttinger was soon to be based there and also on the faculty was Keith Brueckner, who had been at the forefront of developing many-body theories for condensed matter physics (see Figure 45). Brueckner had shown how to use such theories for the interactions of many electrons and these methods would be used by quantum chemists.[195] The diagrammatic methods of Feynman had been extended to many-body theory by Goldstone and Brueckner, and Walter was keen to learn about these new mathematical techniques which had promise for solid-state physics.

Brueckner had recently published a paper in the *Physical Review* which stated: "It is shown for systems of strongly interacting particles that in the limit of very many particles a transformation exists leading to an alternative problem which can be solved by a self-consistent field method."[196] This paper may have helped set the seeds in Walter's mind for the computational method he would develop with Sham for implementing DFT. Brueckner also wrote papers in the 1950s discussing the "exchange-correlation energy", a term which would become central in the implementation of the DFT.[197] Walter's first entry into many-body theory was published in the *Physical Review* and had the title "Effective mass theory in solids from a many-particle standpoint".[198] Sabbaticals can often cause perturbations and this was true in Walter's case from several points of view. He also took another sabbatical at Imperial College London in 1958 with Harry Jones, whom he had tried to recruit previously to Carnegie Tech.

Through the 1950s, Walter was publishing 6–8 papers a year on theoretical solid-state physics, almost all in the *Physical Review*. This productivity was being noticed and, as was the case back in 1953, he started in 1957 to receive again significant approaches, and this time it was for full professorships with tenure. The universities of Chicago, Pennsylvania, Iowa State and Michigan all made Walter offers with 10-month salaries ranging from $12,500 to $14,000. Walter even had a tempting approach from IBM in New York at $17,000 per annum. However, as was the case three years before, President Warner of Carnegie Tech came back with an offer, in this case of $14,000 with full tenure.[199] With his additional Bell Labs summer salary, this was competitive with IBM. Walter, therefore, took the easiest route and stayed put at Carnegie, at least for a little while.

It was during the 1950s that Walter made the acquaintance of Robert ("Bob") Parr. He was a theoretical chemistry professor in the chemistry department at Carnegie Tech. With theoretical chemistry involved with applying quantum mechanics to molecules Walter and Parr had discussions on problems of mutual interest and some of Parr's students had taken Walter's lecture courses. Parr had also acted as an internal examiner for the PhD examinations of some of Walter's students. Together with Rudolf Pariser, Parr had proposed in 1953 an approximate procedure for calculating the molecular orbitals of conjugated molecules.[200] Working in Cambridge, England, John Pople had published a similar method at the same time and it became known as the Pariser-Parr-Pople theory.[201] In due course, Parr was going to be a crucial catalyst for the Nobel Prizes award to Walter and Pople.

In 1956, Ed Creutz, who had been such a strong supporter of Walter, left Carnegie Tech to join the General Atomic Division of General Dynamics in San Diego. This company was to work on peaceful applications of atomic energy. Creutz was recruited by Frederic de Hoffmann, the close friend of Walter from his time at Harvard, with whom, as we have seen, he corresponded. By this time, Walter was on his own two feet at Carnegie Tech and did not need the support of Creutz in the department any more. However, this link was very soon to be re-established when Walter moved to San Diego.

Bell Labs continued to be impressed by Walter, and Conyers Herring wrote to offer him $1,750 for his summer salary in 1959.[202] This was over double what he was getting ten years before. Walter had impressed the management at Bell Labs with his productivity during his summer visits, as Anderson recalled:

> We were able to use him as a stick with which to beat the Bell Labs management, which, if possible, appreciated him even more than we did. In 1955–56 we were encouraged to organize a new theory department, probably because of the rather rapid attrition of theorists from the Labs. We had of course lost Charlie Kittel and John Bardeen ... One of the commonest phrases we used in the negotiations with our managers was "let's organize a department into which Walter Kohn would be happy to come." We managed to get, with this as our strongest argument, a democratic structure with rotating chairmanship, postdocs, of course a measure of control over our own fates (until then we had been separately supervised by experimental department heads), a liberal travel and sabbatical policy, and (quietly and informally) considerably improved salaries — a lot to owe Walter for![203]

While on sabbatical at Imperial College early in 1958, Walter wrote several papers with new ideas. He studied the periodic mathematical functions for solids that Felix Bloch had introduced back in 1929 and examined how they were perturbed by a magnetic field. He also discovered that the phonon (vibrational) spectrum of metals can show anomalies.[204] This effect, now known as "Kohn Anomalies", has been observed more recently in several materials such as graphite and graphene.[205] In addition, in the fall of 1958 James Langer had moved to a position at Carnegie Tech after his PhD on a Marshall Scholarship with Rudolf Peierls in Birmingham, UK. With Walter's postdoc Seymour Vosko he applied perturbation theory to calculate the effect of a small fixed charge on a high-density electron gas. Kohn and Vosko then extended the approach to calculate the nuclear magnetic resonance signal in copper due to the presence of solute atoms.[206] This prediction was soon verified by experiment. By the end of the 1950s, there were few theoretical solid-state physicists who were competing with Walter Kohn.

However, things were not settled for Walter in Pittsburgh for long. Brueckner had been commissioned to start up a new Department of Physics in the beautiful location of La Jolla right on the Pacific Ocean in California. He had been very impressed by Walter's work and his visit to Philadelphia. Accordingly, Brueckner proposed to Walter in 1959 that he join him in La Jolla.

Chapter Eight

San Diego

Roger Revelle was a professor at the Scripps Institution of Oceanography in La Jolla, a community just north of San Diego (see Figure 46).[207] A tall, imposing man he had become prominent in the US scientific community, especially in rapidly growing California. He had chaired a very influential committee of the National Academy of Sciences which communicated the damaging biological effects of atomic radiation. He had also published several early papers on the take-up of greenhouse gases such as carbon dioxide by the oceans and indicating future problems for climate change. A 1957 report on his research was the first to use the phrase "global warming".[207]

The Scripps Institution had been created through a donation from Ellen Scripps, who lived in La Jolla from 1897 until she died in 1932.[208] She had made major donations to local institutions, including the hospital which was in the Scripps name. Her benefaction helped create the Scripps Institute for Biological Research, which became the Scripps Institution of Oceanography of the University of California in 1925. Ellen Clark, the great niece of Ellen

Fig. 46.   Roger Revelle, 1964. (Courtesy of Special Collections & Archives, UC San Diego.)

Scripps, married Revelle. This gave him a unique influence into the La Jolla hierarchy.

Revelle was a highly energetic political operator in Southern California where he was well connected from every point of view. He was a naval officer in the war and he led the University of California Division of War Research in San Diego. There was an increasing demand for science and engineering graduates in the naval defence industry in the San Diego area. Revelle had also served as President of the La Jolla Town Council. In addition, as a member of the Academic Senate of the University of California, Los Angeles (UCLA), he had promoted at an early stage the idea of an expansion of a new campus in San Diego as part of the University of California.[207] This initially received some resistance from influential academics at UCLA being concerned with serious competition close by.

However, with the considerable rise of students wanting places at UCLA and Berkeley, a Master Plan for Higher Education in California, commissioned by the State Board on Education and the Regents of the University of California, recommended the expansion of the University of California by the creation of new campuses.[209] This gave Revelle the opportunity to make the firm case for a new University of California campus at La Jolla. The Regents of the University of California, with strong support from the President Clark Kerr, then authorised the establishment of the campus for graduate research in science, mathematics and engineering with considerable resources for new faculty appointments.[210] Extra financial support from General Atomic was important in this decision.

Revelle, somewhat controversially, also negotiated with the City Council in San Diego for land to be bought for this purpose on a beautiful site at La Jolla overlooking the Pacific Ocean and close to the Scripps Institution of Oceanography. After much heated discussion this was agreed, providing the proposed name was changed from the University of California, La Jolla, to the University of California, San Diego (UCSD).[210] Revelle at once started to recruit leading academics in several subjects. He was quickly successful in chemistry where Jim Arnold from Princeton, a pioneer on carbon dating, and Harold Urey from Chicago, who had won the Nobel Prize for Chemistry in 1934 for the discovery of deuterium, were recruited.[211,212] With the new field of space science rapidly emerging, Urey and Arnold anticipated the opportunity to establish a new research programme in cosmochemistry in San Diego.

These moves to San Diego were treated with disbelief by some of the traditional hierarchy in the venerable academic institutions in the eastern cities in the USA. For example, when Arnold informed his departmental chair at Princeton that he was going to move to San Diego, he was asked: "How could you possibly leave a venerable old university like Princeton for an unknown, proposed new University of California campus?" Arnold replied: "I am not a priest, I'm a missionary."[213] That was the pioneering attitude of the professors who were to join the new campus.

Keith Brueckner, whom Walter had interacted with during his sabbatical at the University of Pennsylvania, was an associate of Frederic de Hoffmann, the friend of Walter's from his graduate days who had founded the company General Atomic in San Diego with Ed Creutz.[142,214,215] Brueckner worked as a consultant with General Atomic on some extraordinary and controversial projects.[214,215] One, called Project Orion, involved examining if a spacecraft could be propelled by a nuclear bomb.[214] During a visit of Brueckner to General Atomic he was introduced to Revelle, who at once asked him to help set up the Department of Physics at what was then called the University of La Jolla. Like many others, Brueckner was inspired by the beautiful site at La Jolla and the exciting, but risky, opportunity to establish his own department. Revelle later commented on Brueckner: "Starting a new physics department, in a non-existent university, in a remote resort town, where he would be surrounded by oceanographers, was just the kind of far-out gamble that he would be completely unable to resist."[215]

Brueckner realised that some research fields would be very expensive, such as particle and nuclear physics. However, solid-state physics, which was developing so quickly in the 1950s, was much less expensive to set up, especially with theoreticians. Since he had recently interacted with Walter Kohn in Philadelphia, he was a prime target to attempt to recruit as a founding physicist at the new university department in La Jolla.[214]

Walter had been at Carnegie Tech for nearly ten years. There had been some ups and downs but overall the Institute had treated him well. He had initially been granted leave for two years, was appointed to an Associate Professorship after only three semesters of teaching and was promoted to a Full Professorship shortly after. However, Brueckner had put to Walter that he would be able to play a key role in recruiting to San Diego colleagues from establishments such as Bell Labs where his connections were first-rate.

In the typical style of leading academics in the USA, Walter then went out of his way to try to attract to La Jolla many of his friends he had met on his academic journey. The contrast of the weather and environment of La Jolla to Pittsburgh was stark. Walter had never worked close to the Pacific Ocean and his letters around this time clearly show he was bowled over by the excitement and change of environment. However, this was a risky move. There was, as yet, no department of physics, no colleagues (apart from Brueckner), no students, and no suitable building in La Jolla. Just two years before, Walter had received offers of full professorships from major schools, including Chicago and Pennsylvania, and he probably anticipated that if his appointment in La Jolla did not work well he would have other more traditional places to move to.

During 1959 Walter made several visits to La Jolla to talk to Revelle and Brueckner, and Lois also came out in September. As early as May he was sent a telegram by Revelle with the offer of a salary of $16,000 for nine months.[216] This was $2,000 more than he was getting at Carnegie Tech. Revelle then sent a telegram to Walter on 19 July confirming the offer from the Board of Regents, University of California.[217] Walter agreed he would come but with one condition — that Brueckner became Chair of the new Department of Physics. After further negotiations over details including housing and starting date, Walter wrote on 22 October 1959 to John Warner, a chemistry professor and president of Carnegie Tech, to resign his post:

Ever since I came here in 1950, my association with Carnegie Tech has been a most happy one from my point of view. In particular the Physics Department is very fortunate in having a staff of unusually fine people who create an excellent atmosphere for academic work, which is difficult to find elsewhere. May I also say at this time how much I have appreciated the liberality with which you and your administration have given me opportunities for spending extended periods at other institutions.

> We shall be leaving Carnegie Tech and Pittsburgh with much sadness. Our reasons for deciding to go to California are largely of a personal nature: The opportunity of participating in the development of a new and promising institution, our hope of finding a somewhat better high school for our girls than here in Pittsburgh, the unusual beauty of the location.[218]

With Carnegie Tech having made a special effort to retain Walter over a period of ten years, Warner responded in somewhat negative terms: "I am especially disappointed when one of our first-rate people decides to go elsewhere."[219] Ed Creutz had also written to confirm Walter's continuing consulting role with General Atomic, very conveniently in La Jolla.[220] Walter was keen to recruit his best co-workers and colleagues to appointments in the new department and was given encouragement from Brueckner to do this. This included Luttinger and Jost. Luttinger travelled to La Jolla in September 1959. Brueckner, a former colleague of Luttinger in Philadelphia, reported to Walter that the visit went well but the New Yorker "found the social life here not up to his hopes".[221]

Another possibility was Ben Mottelson who had overlapped with Walter in Schwinger's group at Harvard and also in Bohr's Institute in Copenhagen. Mottelson had remained in Copenhagen collaborating with Aage Bohr on the energy levels of atomic nuclei, work which would win them both the Nobel Prize for Physics in 1975. At that time in 1959, Mottelson was on a sabbatical at Berkeley, a university which had shown an interest in appointing him. Walter wrote a detailed letter explaining his motivations for moving to La Jolla and encouraging Mottelson to come too:

> Let me perhaps begin by describing how I reached the decision to leave here and go to La Jolla. As you probably guess, I had initially considerable misgivings about moving to such a material paradise which does not yet have those

human and cultural qualities which are very important to us. But after visiting there, seeing the progress being made, the enthusiasm for the future, the strong support from the city of San Diego and the President and Regents of the University, I reached the conclusion that here was an exciting and unique opportunity for making — according to one's own ideas — a real contribution to the scientific and educational life of the country. Of course joining a new organization such as this involves a certain risk... And, to put it very simply, I felt that if I declined I would ever after feel that I had acted without enterprise, putting a very satisfactory certainty ahead of wonderful new possibilities.[222]

Walter then went on to extol the advantages of the environment of La Jolla and gave some negative comments on Berkeley:

So much for purely professional considerations. Another mention was how the family would take to living there. The God-given advantages of the place are of course obvious and as we all love the outdoors this was an important positive factor... The city of San Diego had a referendum on whether to offer the University a large (1,200 acres) tract of public land in La Jolla for the new campus. The vote was decidedly in favour... La Jolla has fundamentally a very strong sympathy with the idea of having a University community in their midst, something similar to Princeton perhaps... I realize that Berkeley may give us some competition — they certainly have a remarkable faculty — but frankly I find it hard to visualize you as a Berkeley man. I don't know a single reasonable person who has been or is on that faculty without being pained by the prevalent concept of science and by the deep divisions existing there...

> By the way we are moving out in January and you may be interested in what our plans for housing are… One very attractive possibility is to rent or buy a house from a non-profit organization which has been set up by some Scripps people for the benefit of new faculty. They have already bought some very nice (and quite reasonable) houses and are about to buy a few more.[222]

Despite Walter's very persuasive efforts, Luttinger, Jost and Mottelson did not follow Walter to La Jolla. Luttinger was soon offered a prestigious professorship at Columbia University in his home city of New York, Mottelson went back to Copenhagen and Jost had just been appointed to a full professorship at the ETH, Zurich. However, Walter was not to be discouraged. He now turned to his many colleagues at Bell Labs and here he had considerable success. Harry Suhl, an expert in statistical mechanics, George Feher, then working on magnetic spin resonance, and Bernd Matthias, a brilliant superconductivity experimentalist, all agreed to join the new Department of Physics, even if setting up suitable laboratories was still a challenge to be met. A subsequent comment from Revelle perhaps explains why the recruitment from Bell Labs was so successful:

> We paid them what they were getting at the Bell Labs, but I was so naive that I didn't realize that just about doubled their incomes because here they were paid on a nine-month basis, and they were paid on a twelve-month basis at Bell Labs. Here they could consult in the summer time, and they were so good that they could get fabulous salaries in the summer. So they got twice as much here as they were getting at Bell Labs… I wouldn't say it was stupid on my part, but it was naive on my part not to realize this.[223]

The alert commentator Anderson, however, was not so happy about Walter's recruiting role:

Walter Kohn went to La Jolla, became the department chairman, and immediately started hiring away all of our best scientists from Bell Labs. And it wasn't long before he had hired away Bernd Matthias and George Feher and Harry Suhl. He had the beaches of La Jolla and the marvelous climate and the marvelous atmosphere. At that time it was a much smaller university and you had a feeling of an expansive world. And you had Nobel Prize winners like Harold Urey and future Nobel Prize winners like Maria Mayer and Joe Mayer that were a part of the social scene. It was very glamorous and a rather fast social scene, it turned out. Very different from the puritanical and Victorian-type social scene that we had at Bell Laboratories.[224]

Larry Peterson was a space physicist who was recruited around this time. His experience of coming from the mid-west to San Diego in the winter was typical:

I left Minneapolis in a blizzard situation and landed in warm and sunny San Diego, welcomed by the founding Chair of the emerging Physics Department, Keith Brueckner, dressed in a short sleeve shirt, shorts and sandals! He took me to the Scripps Institute of Oceanography, where the Physics Department was then located, and set me on a bench to see the surf and the sand of the ocean shore. The decision was easy![225]

In the midst of his negotiations, Walter was approached by the famous physicist Edward Condon. In Germany in the 1920s, he had worked with Born, Sommerfeld and Franck in the early days of quantum mechanics and was responsible for the widely applied Franck-Condon principle. After the war he was appointed Director of the National Bureau of Standards. Condon stated that his University of Washington at St Louis wanted to appoint a leading theoretical solid-state physicist and asked for Walter's advice, perhaps hoping he would suggest himself.[226] Walter listed Seitz, Bardeen, Fröhlich, Anderson, Overhauser, Nozières and Schrieffer, which indicates who he rated at the top of his field.[227]

The word was getting round about the enterprising new department in La Jolla at the new university at San Diego and enquires started to come in. Elliott Lieb was one example.[228] A recent PhD from the group of Peierls in Birmingham, he was highly rated by Walter and was subsequently to prove a fundamental theorem underlying DFT.

One of the most bitter areas of controversy about the new university at La Jolla was anti-Semitism. There had been discrimination against Jewish people buying houses in La Jolla for many years and this continued in the period when the new university was being established. This started to be a difficulty for some of the faculty members who were the first arrivals and Revelle publicly expressed his concerns. This suddenly made him unpopular in the area.[229] When Walter had been negotiating with McGill University in Canada some seven years before, anti-Semitism in Montreal was a major concern for him, as described in Chapter Seven.[174] However, before their arrival, Revelle had assured Walter and his colleagues that anti-Semitism was a thing of the past in La Jolla. To test this out, Walter's Bell Labs colleagues George Feher, Harry Suhl and Bernd Matthias had put Walter up for the La Jolla beach and tennis clubs and there was no problem in his membership.[229]

The astute Brueckner and Revelle had noted the possibility of appointing distinguished scientific couples to the new UCSD department. This included Maria Goeppert Mayer and her physical chemist husband Joe Mayer. They had come from Chicago with the recommendation of Urey and this was, remarkably, her first full professorial appointment. She had been a student in the famous research group of Max Born in Göttingen in the late 1920s when she had discovered two-photon absorption by atoms. The timing of her appointment to UCSD was perfect as she was very soon to win the Nobel Prize for Physics in 1963 for her nuclear shell model.[230] She was only the second woman, after Marie Curie, to win this Prize. Other major acquisitions included the distinguished British astrophysicists Margaret and Geoffrey

Burbidge. They had been at Pasadena at nearby Caltech where she had also suffered from gender discrimination.[231]

The small faculty worked hard to create their graduate school of physics. Early in 1960 a general announcement was sent out with the intention of recruiting graduate students:

In the fall of 1960 the School of Science and Engineering of The University of California, La Jolla will initiate a broad program of graduate study in physics. The chief areas of interest of the present faculty are: 1. Physics of elementary particles. 2. Nuclear forces and structure. 3. Physics of the solid and liquid state. 4. Theoretical biophysics. As the faculty is increased other areas of research will be added. Within the next two years it is foreseen that these will include the physics of plasmas, of the upper atmosphere and of space phenomena.

The present faculty and their interests are: Professor Keith A. Brueckner: Physics of elementary particles, structure of many-body systems including nuclear structure, theory of metals, properties of liquid helium. Professor Walter Elsasser: Noted for his work in geophysics and magneto-hydrodynamics, has recently published a book The Physical Foundation of Biology. At present working with his students in theoretical biophysics. Professor George Feher: Electron spin resonance applied to the exploration of the structure of solids and to the production of polarized nuclei. Professor Walter Kohn: Theory of solid-state with emphasis on electronic properties of semi-conductors, metals, and alloys. Professor Leonard N. Liebermann: Supersonic propagation of underwater sound, molecular and chemical physics. Professor Bernd T. Matthias: Experimental solid-state physics including superconductivity, ferromagnetism, ferroelectricity. Professor Maria Goeppert Mayer: Low energy nuclear physics with special emphasis on the theory of nuclear structure.[232]

Fig. 47.   Aerial view of the early campus in La Jolla, 1960. (Courtesy of Special Collections & Archives, UC San Diego.)

The announcement included a dramatic aerial photograph of the small campus at La Jolla (see Figure 47).[232] This picture catches the pioneering and isolated atmosphere next to the ocean and also the opportunities for expansion. This was now the 1960s in California and the atmosphere was informal. Students even turned up for lectures dressed in bathing suits.[233]

Walter worked hard to recruit graduate students to the new department and wrote to colleagues in a number of universities in the USA asking if he could visit to explain the virtues of his new department and university. These approaches were not always successful. For example, he wrote to the University of Chicago which had many undergraduates taking courses in physics but they wrote back to say that they were not so keen for Walter to visit.[234] They had lost Urey and the Mayers to San Diego and they may have been worried that Walter would recruit even more leading physicists in the same way that had been the case with Bell Labs. He wrote a similar letter to Eugene Wigner who replied to say that all the Princeton undergraduate students had already decided on their graduate schools.[235] Walter continued to suggest international names for appointments, sometimes a little naively. For example, in one letter to Brueckner he suggested several physicists from the University of Cambridge in the UK including Heine, Kemmer and Polkinghorne and also Friedel and Aigrain from France.[236] None of these were recruited.

The recruitment drive for graduate students was quite successful and, by the fall of 1960, as many as 16 had started in the new Department of Physics. This number was doubled by 1963.[233] Walter had the immediate benefit of students and postdocs he brought from Pittsburgh. This included Edwin Woll, Hiroshi Hasegawa, Emile Daniel, Anthony Houghton and Seymour Vosko. A new postdoc, Stephen Nettel, also joined. As Walter's research group was purely theoretical with minimal computational facilities, all they needed was accommodation and desk space.

***

On 1 January 1960 Walter had a letter in German from an aunt Lily in Lyon, France about his uncle Joseph ("Jonny") Rapaport who had nearly died of heart problems.[237] With very few of the older generation of Rapaports surviving the Nazis, Walter was always keen to see these members of his family when he could. He was particularly close to his uncle Joseph, the brother of his mother Gittel (see Figure 48, and also Figure 1 for Joseph as a young boy). It was therefore important for him to go to France and, as we shall see, his next visit there would be a crucial one for his research.

Almost as soon as he started in La Jolla in January 1960 Walter was writing to Jost extolling the virtues of his new environment:

> We have now arrived at the shores of the Pacific and are still trying to get used to the breath-taking beauty of this place... We would love to have you visit us here for any length of time that you can spare, preferably years! And since this is such a wonderful, relaxing place you would almost have to bring your family with you. Just mention to Hilde that we have already been swimming in the ocean several times and see what her reaction is![238]

Fig. 48. Lily Mueller Rapaport and Joseph ("Jonny") Rapaport, the aunt and uncle of Walter Kohn. (Courtesy of the family of Walter Kohn.)

In 1960 Walter heard he had won the Oliver E. Buckley Prize for Solid-State Physics of the American Physical Society.[239] This was his first major prize. It was awarded for "his extension and elucidation of the foundations of the electron theory of solids" and pre-dated his work on DFT. He was always particularly proud of this award as the first winners were his Nobel Laureate colleagues Shockley and Bardeen from Bell Labs. He was awarded the prestigious prize three years before Anderson, always a friendly competitor with Walter.

The move to La Jolla had an immediate impact on the research productivity of Walter and his group. During 1959 and 1960 he had published as many as 12 papers from the Carnegie Tech address while in the three years of 1961 to 1963 the number from UCSD was just four. One paper published in the *Physical Review* in 1960 was with Nettel on "Giant fluctuations in a degenerate Fermi gas".[240] This paper showed his ideas were now heading towards electron densities.

Walter had agreed to come to La Jolla provided Brueckner was Chair of the new Department of Physics. However, in 1961, after just one year in the role, Brueckner took leave from the university for two years to become Vice-President of the Institute for Defense Analyses taking over from Charles Townes.[214] This Institute had a steering committee called JASON consisting of some of the leading physicists in the USA including Murray Gell-Mann and Freeman Dyson. The group became controversial and "would not be asked specific questions by government agencies, but rather be presented with a variety of issues and concerns out of which they would formulate questions for deeper investigation."[214] After Brueckner, Walter had played a more active role than anyone in recruiting new faculty and he reluctantly took over the role of departmental Chair.

Walter started to become involved with political academic controversies which may have distracted his research. Alarmed by continuing uncertainties about the land allocated to his new university, he wrote a stern letter in June 1960 to the Governor of the State of California Edmund Brown:

I am a new faculty member (Professor of Physics) at the University of California, School of Science and Engineering, in La Jolla. From recent newspaper reports, I gather that because of the uncertain future of Camp Matthews the Regents may not accept the land offered to the University by the City of San Diego, and thus may put into very serious doubt the establishment of a general university campus in the La Jolla area. I also understand that you have called a meeting for June 23 to discuss the State's plans for higher education. In this situation I regard it as my responsibility to explain to you my considered opinion — shared by many of my colleagues — that unless a positive decision is made concerning a general campus here, a unique opportunity for building one of the world's great universities would be seriously jeopardized and probably lost.[241]

The letter did seem to have some effect and Governor Brown wrote back:

Thank you for your letter supporting the development of a general university campus in the La Jolla area. We have every desire in this state to develop fully the facilities for all educational activities, and, of course, for the University. We have just had a very constructive meeting in which the problem of developing new campuses was discussed along with the financial potential of the State. I appreciate your interest and concern in accelerating the activation of the La Jolla Campus.[242]

This was the start of several campaigns that Walter would be involved with in academic politics, essentially for the rest of his career. In the 1930s in Vienna, he had seen at first hand the hugely negative influence politics can have on education. In the USA, when he was upset about a political aspect that came into his sphere, he did not hesitate to express his view.

Walter's letter to the State Governor relates to the major effort of Revelle to bring Jonas Salk to La Jolla. Salk had become world-famous for his development of a vaccine for polio and the authorities in La Jolla were very keen for him to come. He had carried out this work at the University of Pittsburgh School of Medicine and had met Walter there. At La Jolla, he had the opportunity to establish his own institute but there was a disagreement over the land allocated for this purpose and whether this would take away some of the area which had been agreed for the new university. Walter and Salk swapped some quite sharply worded letters on this issue. Salk wrote to Walter to clear the air:

I deeply appreciate the sincerity of your very frank letter and the spirit in which it was written. In view of what has transpired in the past few weeks, a number of things are under consideration; I am not one who believes in miracles. It will take some time for the issues to become clarified. Under no circumstances would I want to do anything that would jeopardize your "own plans for building a great university here."[243]

The local council put the matter up for a vote which was won in Salk's favour and Walter wrote to Salk to congratulate him. Then there was a huge row in February 1961 about the appointment of the first Chancellor of UCSD. Revelle was the obvious candidate and he wanted the job. He was strongly supported by the newly appointed faculty including Walter. However, he had made himself unpopular with the local Council through his strong pushing for the new university, the problems with the land for the Salk Institute and his concern on anti-Semitism issues over housing. This weighed heavily with the Regents of the University of California.

Herbert York, a nuclear physicist who had worked in the Manhattan Project and was formerly Director of the Radiation Laboratory at Lawrence Livermore, was appointed Chancellor. The hugely disappointed Revelle left UCSD in 1963 and founded a Centre for Population Studies at Harvard.[207] After 13 years there things had cooled down and he returned to UCSD in 1976 as a Professor of Science, Technology and Public Affairs. He would win the prestigious Balzan Prize and the National Medal of Science for being, in his own words, "the grandfather of the greenhouse effect".[207]

Revelle was always very proud of the University he had created at La Jolla almost from scratch. He was delighted with the many awards his appointments received. He listed Keith Brueckner, Walter Kohn, Bernd Matthias, George Feher, Margaret Burbidge, Geoffrey Burbidge, Norman Kroll, Oreste Piccioni, Harry Suhl, and Marshall Rosenbluth as physicists who would be elected to the National Academy of Sciences (or its British equivalent the Royal Society) during their time at La Jolla.[223] In addition, Maria Goeppert Mayer, already a member of the Academy, would win the Nobel Prize for Physics while in San Diego and George Feher would be awarded the Wolf Prize for Chemistry for his work in biophysics. Furthermore, Walter would win the Nobel Prize for Chemistry for research he did while he was a member of the Department of Physics at UCSD.

Walter was pleased to hand over the headship of the Department of Physics to Norman Kroll, whom he had been involved in appointing. This was quite a coup as Kroll came from a full professorship at Columbia University. He had done early work on quantum electrodynamics with Willis Lamb. Walter had also appointed his former collaborator Norman Rostoker, who had moved into plasma physics.

The third daughter to Walter and Lois was born in November 1962. She was given the name Eva Rosalind, and Walter commented that the first name was chosen from Eva Hauff who had been so kind to Walter and his sister in England in 1939 and 1940.[6] After the war, Eva and Charles Hauff left their house in Copthorne, Sussex to move to Ilkley in Yorkshire. Charles died there in 1955 and Eva in 1959. Their house, the Border Cottage in Copthorne, where Walter had lived during 1939 and 1940, was replaced by a new residential road, the Border Chase, in the 1970s and 80s.

***

With all the administration and politics hindering his research in the early 1960s, Walter needed a break and another sabbatical in Europe was his ideal. He was successful in an application to the Guggenheim Foundation for a Fellowship in the fall of 1963 at the École Normale Supérieure in Paris. His stated topic was "Interaction of electrons and phonons with metals".[54] He anticipated he would have good interactions there with Jacques Friedel and Philippe Nozières. He would also be able to see, perhaps for the last time, his uncle in Lyon, the most senior member of his old family in Europe. In addition, he enjoyed the artistic treasures of Paris, especially the medieval tapestries in the Musée de Cluny.

In June 1963, Walter received a welcome letter from Clark Kerr, President of the University of California, with an increase in his salary to $19,200.[244] This was an enhancement of 20% at a time of low inflation. He had worked very hard to make the new Department of Physics at UCSD a success and this had been recognised at the top of the University of California. That year Walter was also elected to membership of the American Academy of Arts and Sciences, one of the oldest academic societies in the USA founded by John Adams and colleagues in 1780.

In the spring of 1963 Walter had written a paper on the theory of insulators.[245] This would become one of his more highly cited publications. In October of that year Eugene Wigner was awarded the Nobel Prize for Physics, mainly for his work on "fundamental symmetry principles". Walter had written him a congratulatory letter and Wigner replied saying that Walter's new paper on insulating states "was a must not only for solid-state physicists but to all physicists".[246] Walter had achieved a huge amount in his forty years despite the extraordinary obstacles which had been put in his

Fig. 49. Pierre Hohenberg, 1982. (Courtesy of American Institute of Physics, Emilio Segrè Visual Archives, Physics Today Collection.)

way in his early days. He was now regarded as one of the leading theoretical solid-state physicists. However, he had not yet published an iconic work that would deserve to make him a household name in science. That situation would soon change in a major way.

At the École Normale Supérieure, Walter's host Philippe Nozières had just published jointly with Luttinger.[247] Nozières had a new postdoc Pierre Hohenberg (see Figure 49). He had a PhD from Harvard where he worked on superfluid helium with Paul Martin.[248] He had just spent a year in Moscow where he published a paper on superconductors with non-magnetic impurities.[249] His father was born and raised in Vienna before moving to Paris, and this gave a special bond with Walter. Hohenberg wrote:

> I first met Walter Kohn in 1963 when I was a fresh Ph.D. spending a post-doc year in Paris at the École Normale Supérieure in the group of Philippe Nozières. As it happened, Walter was also spending time at the École Normale and conditions being what they were I was privileged to share an

office with him for an extended period. If I remember correctly, the relatively large space was Philippe's office and even if it wasn't I remember it to have been a general meeting place and thoroughfare, a little bit like trying to think deep thoughts in the middle of Times Square.[2]

Walter arrived in Paris at the beginning of September 1963 after a brief visit to see his old friends in Copenhagen. Sadly, there was now no Niels Bohr, who had died the year before at the age of 77. Walter wrote about how his own thoughts developed in Paris:

I had been interested in the electronic structure of alloys, a subject of intense experimental interest in both the physics and metallurgy departments… The question occurred to me whether, in general, an alloy is completely or only partially characterized by its electron density distribution… It occurred to me that for a single particle there is an explicit elementary relation between the potential and the density of the ground state. Taken together, these provided strong support for the conjecture that the density completely determines the external potential … and all properties derivable from [the Hamiltonian] H. It turned out that a simple 3-line argument, using my beloved Rayleigh-Ritz variational principle, confirmed the conjecture.[11]

In the back of his mind, Walter may also have been recalling the analogy to his work with Jost on quantum collision theory when they derived a rigorous method for determining the potential energy from scattering data.[163] Walter needed confirmation that his simple idea on the electron density was correct. He had found with Luttinger and Jost that his most decisive research was done in close collaboration with a co-worker and he identified in Hohenberg just the right characteristic of a strong background in mathematical physics. Furthermore, they were sharing a desk in Philippe Nozière's office.

Walter suggested that Hohenberg survey the literature to see if his idea on electron density was already known, but nothing could be found. With Hohenberg they "fleshed out this work with various approximations and published it".[250] Their paper contained two original results. The first was that the ground-state electron density uniquely determines the energy, electronic wavefunction and all properties of the system in question. The second fundamental result was that the functional for the electronic energy is a minimum for the exact ground-state density. Little did they know that this foundation paper on DFT, entitled rather obscurely "Inhomogeneous electron gas", would become for some time the most highly cited paper in the *Physical Review*.[251]

Walter and Hohenberg shared their ideas with colleagues in France but did not initially find acceptance. Nozières said: "In my opinion, putting all the emphasis on the density did not account properly for exchange and correlations. I did not share the enthusiasms of Walter and Pierre and I stayed aside."[54] Nozières was correct in identifying the difficulties with exchange (the principle that a wavefunction must change sign when any two electrons are interchanged) and correlation (arising from the Coulomb repulsion between electrons). Exchange and correlation would be key issues in DFT that would take many years to be addressed to make the theory generally useful across the sciences.

In December 1963 Walter made his first visit to Russia, attending a conference in Moscow on solid-state theory with his new co-worker Hohenberg. This would initiate an interest in Russian scientists that would last for the rest of his career. Then in January 1964 he went to Cambridge, at the invitation of Nevill Mott (see Figure 50), and Oxford where Rudolf Peierls was now based. Mott had been appointed Cavendish Professor at Cambridge University in 1954, replacing Lawrence Bragg. This was the leading physics professorship in the UK — before Bragg the Chair had been held by Rutherford, J. J. Thomson, Rayleigh and Maxwell. Mott's move to Cambridge opened up major research lines in solid-state physics which even attracted Anderson in 1967 to an appointment split between Bell Labs and Cambridge. Mott's interest had paralleled Walter's in that he initially had started research on the quantum theory of atomic collisions and wrote an

Fig. 50. Nevill Mott, 1950. (Courtesy of American Institute of Physics, Emilio Segrè Visual Archives, Physics Today Collection.)

acclaimed textbook on the subject with Harrie Massey.[252] Then in the mid-1930s he moved over to solid-state physics, and, in due course, was to write:

> The impact of solid-state electronics on our society has been enormous. It has given rise to whole new industries, new professions, new ways of production and new organizations. Moreover, unlike many industrial innovations, it depends to an exceptional extent on developments in pure, abstract science.[253]

Walter was sent a letter by Brian Pippard at Cambridge with a warning on his seminar there: "The audience is rather varied, consisting of research students working in a wide range of solid-state problems and more acquainted with experiment than theory in all too many cases, so if you can find it in your heart to give us a simple little chat on general matters rather than a detailed exposition of those nice points for which you are so famous, I think such a talk would be more generally appreciated."[254] This message expresses a common sentiment from experimentalists when a theoretician is coming to give a departmental talk.

Walter had also received in Paris an invitation from the University of Toronto to take up a Visiting Professorship. His alma mater was finally waking up to his international reputation and he agreed to return in the summer of 1964, but just for one week.[255]

After returning to La Jolla after his short but highly productive sabbatical in Paris, Walter continued to put the finishing touches to the paper with Hohenberg. Walter wrote to Hohenberg in Paris on 15 June 1964 just after submitting their paper to the *Physical Review*:

I have finally read your work on the general gradient expansion and found it very nice. For the purposes of the present paper, I thought it was sufficient to refer to it briefly and have mentioned it on page 14, which I have also otherwise rewritten in accordance with your suggestions. A copy is enclosed. All the other corrections, which you also suggested, have already been made and I have just sent the paper to the Physical Review.

I may have mentioned to you that Lu Sham and I have started looking at the situation which one has, for example, in the heavy atom where there is a localized, rapidly varying potential and a large region where the potential varies smoothly. We are studying this case, at the moment, for a Fermi gas without interactions, the general idea being to describe the effect of the strong localized potential in terms of phase shifts and then continuing the solution in the outside region by the BWK method. Naive as this approach may sound, we have nevertheless not been able to carry it through yet, but I have a strong feeling in my bones that some simple formulas must exist.

If and when this is done, I would like then to think about the same system with interactions included, and this would

Walter refers to the insertion in their paper of a section on the general
gradient expansion of the electron density which would, in due course, be
picked up in an important way by other groups who would further develop
the computational applications of DFT. He also refers to the ongoing progress
being made with Lu Sham. The "Hohenberg-Kohn" paper was received by
the *Physical Review* on 18 June 1964 and was published on 9 November
1964.[251] It ranks as one of the most influential scientific papers published in
the second half of the 20th century.

Hohenberg then applied for several jobs in the USA. In his recommendation
letters for Hohenberg, Walter included the intriguing comment: "His own
style is more of an artist than a systematic scientist, but he is technically very
capable, and as you can see from some of his work he is able to carry through
quite complex theory."[257] Hohenberg was appointed at Bell Labs, with Walter's
letter having considerable weight. Despite the suggestions in Walter's letter
of 15 June, Hohenberg and Walter did not collaborate on research again.
Hohenberg turned his interests to fundamental theoretical aspects of fluid
dynamics and liquid helium, and his work in these areas also become highly
cited. He was elected to the National Academy of Sciences and, after a period
at Yale University following Bell Labs, he became Senior Vice-Provost for
Research at New York University.

Back in La Jolla, Walter was realising that a computational method was
needed to determine the electron density, in analogy to the methods in use
for determining wavefunctions for systems with many electrons. Despite
being a rather pure theoretician he always had a real instinct for developing
theories that could be implemented quite generally in computations.
A special feature of DFT was that the electron density just depended on three
Cartesian coordinates x, y and z whereas, in normal quantum mechanics,
the wavefunctions had to be determined for electrons each with their own
Cartesian coordinates. It was this multi-dimensional problem that had been

Fig. 51.   Lu Sham, 1972. (Courtesy of Special Collections & Archives, UC San Diego.)

a major feature in preventing conventional wave mechanics to be extended accurately in computations on systems, such as metals and proteins, with large numbers of electrons.

Sham had been waiting in La Jolla for Walter's return (see Figure 51). He had not met Walter until April 1964. He was funded as a postdoctoral assistant on a grant that Walter had obtained from the Office of Naval Research.[258] Sham was born and brought up in Hong Kong and went to Imperial College London for a BSc in Physics. In 1963 he completed his PhD on the theory of electron-phonon interactions at the University of Cambridge working with John Ziman and funded by a Churchill scholarship. At Cambridge he had also interacted with Anderson, who then had a regular visiting appointment there, and with Nevill Mott. On arriving at La Jolla, while Walter was in Paris, Sham wrote a single-author paper for the *Proceedings of the Royal Society*.[259] This was on the phonon spectrum of metallic sodium which reported calculations using the Hartree-Fock self-consistent field method in which the wavefunction is expressed as a determinant containing one-electron orbitals that automatically satisfies the exchange rule for the electrons.

This was to be valuable experience for Sham's next project and Walter suggested that he look at the possibility of developing a computational algorithm to calculate electron densities. In the Hartree method an electron moves independently in an effective potential and Kohn and Sham were able to adapt this procedure for the DFT. They wrote the energy as a sum of a

Hartree term and an exchange-correlation term that both depended on the electron density. They cleverly replaced the interacting system of electrons with a non-interacting reference set which has exactly the same electron density. In this way they derived a set of "Kohn-Sham" equations which are similar to the Hartree form, except that the eigenfunction and eigenvalue solutions had no direct physical meaning but the eigenfunctions are very easily converted into the electron density.

On 13 October 1964 Walter wrote an understated note to David Mermin about this work which had been done over the previous five months: "Lu Sham and I have just finished a little study of one-dimensional inhomogeneous systems and got some rather amusing and promising results, so we are pushing on with the necessary generalizations to three-dimensional and interacting systems and have not yet run into insurmountable difficulties."[260]

Walter and Sham wrote up their new method as a brief report with the title "Exchange and correlation effects in an inhomogeneous gas" which they submitted to *Physical Review Letters* in May 1965. The Editor Samuel Goudsmit, however, replied: "Our judgement is that while it deserves publication, it is not of such urgency to warrant the speedy publication of a letter."[54] Walter and Sham took advantage of this suggestion by revising their paper to include many more details. They gave their revised paper the more general title "Self-consistent equations including exchange and correlation effects" and submitted it to the *Physical Review*. It was received by the Editor on 21 June 1965 and published on 15 November 1965. This is the Kohn-Sham paper that brought the Hohenberg-Kohn theorems to practical calculations on atoms, molecules and solids. It is the second major paper on DFT from Walter and his colleagues that has received a huge number of citations.[261]

Kohn and Sham also proposed a Local Density Approximation (LDA) approximation for the small but crucial exchange-correlation energy. This was done by means of a fit to calculations of the density of an electron gas. In due course, accurate results for an electron gas were obtained with a Quantum Monte Carlo procedure which involves taking the time-dependent Schrödinger equation, making time imaginary and then solving the resulting equation, which resembles that for Brownian motion, with a random-walk algorithm.[262] For several years, the application of the LDA enabled DFT calculations to produce useful results for some solid-state problems. In due course, the LDA would be replaced in many applications by more elaborate exchange-correlation energy terms as described in Chapter Nine. In their paper Kohn and Sham, somewhat ironically, stated: "We do not expect an accurate description of chemical binding."[261] On this point, the quantum chemists were, in due course, to prove them wrong.

Amongst all this busy work, Walter had been able to attend a Gordon Conference at the Kimball Academy in New Hampshire in August 1964 on "Electron-Phonon Interactions" in the series Chemistry and Physics of Solids. This was chaired by John Hopfield who was to later pioneer the theory of neural networks for use in computer simulations and machine learning. Hopfield had invited Walter with the enticing comment: "I realize that the conference attendees will learn far more from you than anyone else on this subject."[263] Accordingly, Walter did not talk on DFT but on "Kohn Anomalies". Van Vleck and Schrieffer were also speakers at the meeting (see Figure 52).

A student in Walter's group, Bok Yin Tong, together with Sham, applied the Kohn-Sham LDA method to several atoms and the results for total energies looked quite promising.[264] Sham then moved into other areas of theoretical physics and took up an assistant professorship at the University of California, Irvine in 1966. Just two years later he was appointed back at UCSD as an Associate Professor. After this, he never left La Jolla and eventually became Dean. The subsequent progress in making DFT a practical and applicable computational method was largely made by many other research groups around the world, and this is described in more detail in the next chapter.

Fig. 52.   Attendees at the Gordon Conference on Electron-Phonon Interactions in the Chemistry and Physics of Solids Series, Kimball Academy, August 1964. From right: 5th John Van Vleck, 7th Walter Kohn and 8th John Hopfield. (Photography Copyright © 1964 Gordon Research Conferences. Used with permission.)

David Mermin had come from the Peierls group in Birmingham, UK to La Jolla in August 1963, just after Walter had left for Europe. He already had an offer of an Assistant Professorship at Cornell University. However, he had postponed his start at Cornell for a year to go to La Jolla after one of his physics friends there had told him: "Volley ball is standard on the beach at noon."[265] Mermin also wrote:

> He [Walter] asked me to think about how to generalize the [Hohenberg-Kohn] theorem from the ground state to thermal equilibrium. I returned to my office to consider it and quickly realised that a strange variational principle for the free energy that I had formulated in Birmingham for an utterly unrelated purpose, seemed to be tailor made for generalizing the Hohenberg-Kohn theorem to non-zero temperature. It took me less than an hour to check that their proof did indeed go through in exactly the same way if the ground state variational principle they used was replaced with my thermal equilibrium variational principle. So I went back to Walter's office and knocked on the door. Here's how you do it, I said. He seemed somewhat taken aback by this … It took me a day to convince him that I had indeed answered his question. Then he was very pleased and I, needless to say, was ecstatic.[266]

After Mermin had returned to Cornell and written up his work for the *Physical Review,* Walter wrote to him on 28 July 1964: "I enjoyed your little note on the Thermal Properties of the Inhomogeneous Electron Gas very much. I think it is exceptionally well done."[267,268,] Mermin's "little note" has now received over 2,000 citations.

Back in La Jolla, Walter felt the time was right to invite the leading theoretical physicists to give seminars and his department finally had its own building (with the exception of a few chemists on one floor). Felix Bloch from Stanford had already visited and now it was the turn of Walter's mentor Van Vleck from Harvard, Bardeen from Illinois and Dyson from Princeton. The UCSD Department of Physics was continuing to grow and

Fig. 53. Faculty of the Department of Physics, UCSD (~1968). Back (from left): Sheldon Schultz, Oreste Piccioni, Ralph Lovberg, Robert Swanson, Shang Kang Ma, John Malmberg, David Wong, Joseph Chen, Herbert York, Frank Halpern, William Frazer, Robert Gould, Thomas O'Neil. Front (from left): Werner Melhop, Xuong Nguyen-Huu, Wayne Vernon, John Wheatley, Walter Kohn, Harry Suhl, Bernd Matthias, William Nierenberg, Barry Block, Margaret Burbidge, Larry Peterson. (Courtesy of Special Collections & Archives, UC San Diego.)

there would soon be over 20 professors with new areas such as space physics and plasma physics (see Figure 53).

Encouraged by the progress of Sham and Mermin with DFT, Walter now gave priority to explaining his new idea in his correspondence and in his lectures. He wrote to Vittorio Celli on 13 January 1965 of the University of Bologna with the news: "Lu and I continue to work on the inhomogeneous electron gas and have some new results which, in our opinion, at least, are not without promise."[269] This was certainly something of an understatement. The Kohn-Sham paper was to accumulate over 50,000 citations in due course.

Walter was invited to give a seminar at Harvard in 1966 and he kept his notes of the talk in his archives.[270] He started off the lecture by stating that although Hylleraas had shown how to include electron correlation accurately in a calculation on the helium atom, there had not been major progress in the next 30 years in accurate calculations on more complicated electronic systems. He described the approximations due to Thomas and Fermi, and that of Fritz London, for treating solids with many electrons. He then went on to carefully explain the basic ideas of the Hohenberg-Kohn paper in which there is a unique electron density from which the properties of the system can be obtained. He also described the Kohn-Sham equations and mentioned the work of Mermin for finite temperatures. He said that ongoing research included calculations on metals and surfaces using the new theory.

Walter had been invited by the distinguished Japanese theoretical physicist Ryogo Kubo to give a lecture in Tokyo in 1965 and write an article for a book he was editing on many-body theory. Walter's paper, which was published in 1966, was entitled "A new formulation of the inhomogeneous electron gas problem" and covered the same ground as for his Harvard lecture.[271]

***

By this time, Walter was getting caught up again with administration and academic politics. He had been Chair of the Department of Physics and that inevitably meant he was vulnerable for requests from the University for committee work. He was made Chair of the Academic Senate of UCSD from 1964–66. He was then involved in a furious row about the library provision in the university. A strong-willed Glaswegian, John Galbraith, who was an imperial historian from UCLA, had been appointed Vice-Chancellor to Herbert York at UCSD, and was quickly made Chancellor in 1964 when York resigned. He made a condition of his appointment that UCSD would expand significantly its library and also its humanities and social sciences programme. However, by 1966 Galbraith felt that he had not received support from Clark Kerr, the President of the University of California, and he resigned in February of that year. As Chair of the Academic Senate, Walter led a campaign to get Galbraith reinstated.[272] This was successful and Galbraith was back in post by the spring. The library was expanded significantly as were the programmes in humanities and social sciences. In due course, the undergraduate library at UCSD was named Galbraith Hall.

Then in 1967 there was a strong political push to remove Clark Kerr from being President of the University of California. This was linked to Ronald

Reagan becoming the new Governor of California. Despite the problems with Galbraith, Walter had a soft spot for Clark Kerr, who had played an important role in the creation of UCSD. Walter was an eloquent voice on the UCSD campus supporting Kerr but this could not stop his resignation.[215]

These developments gave Walter a further taste for academic campaigns which would now frequently occur for him. However, they did detract somewhat from the time he could devote to research. Nevertheless, Walter was still interested in different problems in solid-state physics and, during another sabbatical to Paris in 1967, he published a paper on Mott and Wigner transitions in solids.[273] Mott was impressed and wrote from the Cavendish Laboratory:

> I am rather convinced by your model; it seems to me to allow for a continuous increase of the dielectric constant as one approaches the transition point from the insulator side, and therefore to demolish my original argument that there must be a discontinuous transition. In the light of this I have tried to extend your ideas to a non-crystalline array of atoms and I think this all fits together rather nicely and gives a qualitative explanation of the observations of Fritzsche, Davis and Compton, and so on.[274]

Mott then wrote a series of papers on conduction in non-crystalline systems. His second paper in the series started with the sentence: "The metal-insulator transition for donors in a semiconductor is examined in the light of the work of Kohn on the nature of the transition."[275] Ten years later, Mott was to win the Nobel Prize for Physics, together with Walter's close friends Anderson and Van Vleck, for work on the electronic structure of magnetic and disordered systems.

In 1967, Walter also published with Denis Jérome and Thomas Rice his first attempt at a BCS-type theory for excitonic insulators.[276] In the same year Walter was very pleased to receive a letter from his alma mater the University of Toronto awarding him an Honorary Degree. The President Claude Bissell wrote to say that the Senate was unanimous in recommending Walter for the degree of Doctor of Laws honoris causa because of his "Outstanding contributions to science, and your distinction as a teacher, lecturer and researcher. Our invitation goes out with particular warmth because of our pride in the achievements of a former student."[277] Despite the difficult correspondence with Samuel Beatty some 15 years before, Walter was very pleased. He invited to the ceremony his wife Lois and seven members of her family living in or close to Toronto. It was the first of as many as 18 Honorary Degrees Walter would receive, including from his other alma mater Harvard, Oxford University, Carnegie Mellon University and both the Technical University of Vienna and the University of Vienna. Two years later, in 1969, Walter was elected to the National Academy of Sciences (NAS). He had now truly made it into the US scientific establishment (see his proud gaze in the 1969 photograph of Figure 54).

As he moved towards the 1970s, Walter became more interested in surface science. There was a rapid expansion of high-resolution experimental studies on the interactions of atoms and molecules with surfaces. Walter and his co-workers found that his DFT methods extended quite well to these areas. Neil Lang came as a postdoc in 1967 and worked on metal surfaces. With computational facilities now becoming available, realistic calculations could be carried out by his group, although Walter himself kept to the theory. Walter and Lang wrote two papers on the electronic structure

Fig. 54. Walter Kohn at UCSD, 1969. (Courtesy of Special Collections & Archives, UC San Diego.)

and properties of metal surfaces that quickly received many citations.[278,279] These were the first papers that demonstrated that DFT could provide results that were superior to Hartree theory. Kohn and Lang were awarded the Davisson-Germer Prize for this research. This prize is awarded by the American Physical Society for "outstanding work in atomic physics or surface physics".[280]

Walter and his collaborators did not initially use the term "Density Functional Theory" to describe their method. More laborious phrases such as "Theory for Inhomogeneous Electron Gas" or "Hohenberg-Kohn Theory" were used. However, some papers in the early 1970s then started to use the term Density Functional Theory, initially as a phrase which also covered earlier approaches due to Thomas, Fermi, Dirac and Slater (for further discussion, see the next chapter).[281] In due course, DFT became the catchy acronym for the Hohenberg-Kohn theory despite it also being used in other totally different contexts such as "Discrete Fourier Transform" and "Department for Transport".

A major debate emerged as to whether the DFT method could deal with van der Waals forces and this has continued for almost 50 years. Several of Walter's papers published jointly with members of his research group in the early 1970s were in this area. Nevertheless, he continued to publish a few single-author papers related to topics he had studied before in solid-state physics, including the use of Wannier functions and the very active area of superconductivity.

John Rehr, who collaborated with Walter on Wannier functions, said that, in the early 1970s, UCSD was a "mecca for condensed matter physicists and there were many famous pilgrims".[282] John Bardeen, who had won his second Nobel Prize in 1972 for superconductivity, was one and Alexei Abrikosov from Moscow was another. He was to win the Nobel Prize for Physics in 2003 with Anthony Leggett and Vitaly Ginzburg for theories of condensed matter at very low temperatures. It seems that nearly everybody who visited Walter had won, or would win, the Nobel Prize for Physics. Yet he, himself, was an exception to this rule.

Chapter Nine

DFT

Theoretical chemists had been slow to pick up DFT after the publication of the first papers from Walter and co-workers, but major conferences would help to change that. The first key one was the American Conference on Theoretical Chemistry in Boulder, Colorado in June 1975, on which Bob Parr wrote:

> I do not recall when I first heard of the Hohenberg-Kohn-Sham papers, but I do know that the quantum chemistry community at first paid little attention to them... Walter Kohn's appearance at the Boulder Theoretical Chemistry Conference of 1975 was memorable. On June 24 he presented a formal talk, in which he outlined DFT to the assemblage of skeptical chemists. There were many sharp questions and a shortage of time, so the chair of the conference decided to schedule a special session for the afternoon of June 26. With quite a crowd for an informal extra session like this, Walter held forth on his proof. In his hand he held a reprint of the HK paper, from which he quietly read... The audience sobered down quickly. It was a triumph. The interest of quantum chemists in DFT began to grow at about this time.[283]

Parr, who would be a major catalyst for the Nobel Prize in Chemistry awarded to Pople and Kohn, had the rare advantage over other theoretical chemists that he had got to know Walter quite well at Carnegie Tech in the 1950s (see Figure 55). He was well aware that Walter was a theoretical physicist with fundamental new ideas on quantum mechanics which he developed with mathematical rigour. Following the Boulder conference, Parr developed some DFT applications using "natural" orbitals, which were

familiar in quantum chemistry. Walter invited him to come and speak on the topic in La Jolla and recalled:

> Bob Parr came as a visitor to the University of California, San Diego to give a colloquium on this theory. I had not been aware that he had caught the bug and was wondering what he was going to say. I remember Bob remarking in his introduction that he felt like "carrying coals to Newcastle", but of course, he wasn't doing that at all. As was typical of him, his work was totally original and none of us in San Diego would have ever dreamt of going in that direction. My other recollection from that colloquium is how much I enjoyed having the subject illuminated by a leading theoretical chemist whose perspective reflected that background and was very different from my own.
>
> In those early years of DFT the community of theoretical chemists felt, almost without exception, that this approach had nothing useful to offer to them. Occasionally I was invited to give a paper at their meetings, but I had the feeling that most of the audience expected to confirm their conviction that it was full of irremediable defects, in particular, insufficient

accuracy and the absence of guaranteed, systematic procedures to improve it. The most notable exception was Bob Parr. Secure in his prominent standing in his own community, based on earlier, excellent work in theoretical chemistry, he had convinced himself of the great promise of DFT, and probed for chemically interesting uses in several directions.[284]

Walter was also impressed that Parr was bringing new concepts on electronic structure out of DFT, when the emphasis from other theoretical chemists was on computational techniques. Walter said: "He brought some very fruitful new ways of thinking about chemical reactions, which he sees as consisting as transfers of electron density from one place to another." [284]

As can be seen from the comments of Parr and Walter, theoretical chemists were initially sceptical about the general use of DFT for molecules. An early paper in a chemistry journal on DFT by Kryachko even shed doubt on the proof of the Hohenberg-Kohn theorem.[285] This instigated a major debate and a call for papers on DFT in the *International Journal of Quantum Chemistry*. Mel Levy and John Perdew, who were both going to be major players in bringing DFT into chemistry, quickly published a paper in the same journal which stated: "Kryachko has questioned the validity of the Hohenberg-Kohn theorem. It is our purpose to come to the defense of this and related existence theorems."[286] Perdew, who became a Professor of Physics at Tulane University, had been a postdoc with Vosko in Toronto, who himself had been a postdoc with Walter. Thus Perdew had an inside track into DFT.

It is relevant to consider the progress that had been made in quantum mechanical calculations on molecules from the 1920s up to the 1970s. Following Schrödinger's remarkable paper on the electron in the hydrogen atom in which wave mechanics was first invented in 1926, Heitler and London quickly published a wavefunction that worked quite well in describing the bonding in the hydrogen molecule.[287] Their form consisted of products of atomic orbitals for electrons centred on the different atoms. This "valence bond" idea was quite quickly taken up by Pauling and others to describe the bonding in other molecules such as methane and benzene. However, a different "molecular orbital" approach, developed initially by Lennard-Jones, Mulliken and Hund in the late 1920s, was the method that was found easier to apply in calculations.[288] This method involved expanding the molecular orbital for each electron as a linear combination of atomic orbitals. By expressing the wavefunction as a Slater Determinant containing these molecular orbitals, together with functions describing the spin of the electron, the Pauli exclusion rule was satisfied. Furthermore, adaptations of Hartree-Fock theory produced an iterative self-consistent way of finding the coefficients of the molecular orbitals.[289] All of this computational machinery was established by the early 1950s.

However, this standard molecular orbital approach did not account for the correlation that arises due to the Coulomb repulsion of the electrons. In accurate calculations on the helium atom, Hylleraas surmounted this problem by including the inter-electronic distance in the wavefunction.[290] This procedure has subsequently been extended to many-electron diatomic molecules but it has not been straightforward to apply it to larger molecules within a variational framework.[291] To deal with electron correlation a more cumbersome "configuration interaction" method was needed in which the wavefunction was expanded as a linear combination of determinants with ground-state molecular orbitals replaced by orbitals for excited states. If the orbital basis set and number of determinants is complete, this method, in principle, yields an exact solution of the Schrödinger equation. This approach, however, is very slowly convergent and requires finding the eigenvalues and eigenfunctions of matrices of dimensions that were too large to deal with for many electronic systems of interest. The remarkable improvements in computing power over the decades has allowed for applications of the configuration interaction technique, and variants of it, for more complicated molecules but in the present day it is still not, for example, a computationally viable approach for molecular problems of biological interest.

Perturbation theory and other methods such as coupled-cluster theory, in which particular pairs of electrons are coupled together systematically, helped to get over some of the difficulties but the accurate treatment of electron correlation using these approaches for larger molecules with reasonable computational expense remains a major challenge to this day.[292] These accurate treatments are needed for calculating the activation energies for chemical reactions which requires taking the difference between two electronic energies — those of the reactants and transition state — which are very close in value. Thus quantum chemistry had not produced accurate and generally applicable procedures for calculating electronic energies of chemical reactions by the time DFT was formulated by Walter Kohn and co-workers.

However, quantum chemists found it hard to accept DFT for several years. They had been used to systematic procedures for improving wavefunctions — under the variational principle, the more complicated the wavefunction, the better the result. These systematic procedures did not seem to be feasible in the early days of DFT. Furthermore, and even worse, there was a crucial term in the Kohn-Sham equations, the exchange-correlation energy, which was unknown and there appeared to be no rigorous way of determining it. Thus DFT seemed to fail the main criteria for a good first-principles *ab initio* theory for the electronic structure of molecules. However, in the 1970s, and right through the 1980s and 1990s, there were gradual improvements in the underlying computational procedures which turned Walter Kohn's DFT into a very powerful method for chemistry and materials science. There were a large number of developments achieved by many researchers and here we will only summarise some of the main advances.

The idea that the electron density can determine electronic properties goes back to the very early days of quantum mechanics. Just one year after Schrödinger invented wave mechanics in 1926, Thomas and Fermi examined the electron densities in atoms which they treated statistically.[293,294] Even Dirac was then involved and he showed how electron exchange could be included in the Thomas-Fermi approach.[295] However, the Thomas-Fermi-Dirac model failed to produce the shell structure for atoms and Edward Teller later proved that the method does not even describe the formation of neutral molecules.[296]

John Slater, initially at MIT and then the University of Florida, extended the Thomas-Fermi-Dirac idea (see Figure 56). He solved the Hartree-Fock equations for the kinetic energy but with the term treating electron exchange approximated by the analogous term in the free electron gas. This approach was developed into what became known as the Xα method.[297] Walter always looked up very much towards Slater as he had read his book on chemical physics when he was a student in the Ripples internment camp in Canada

Fig. 56. John C. Slater. (Courtesy of American Institute of Physics, Emilio Segrè Visual Archives, Physics Today Collection.)

in the early 1940s.[65] There were clear connections between the Kohn-Sham procedure and Slater's $X\alpha$ approach. However, it seems that Slater, who died in 1976, was not impressed with the Hohenberg-Kohn approach to DFT which he described as "obvious".[298] It is notable also that Wilson at Harvard, whose theoretical chemistry group had interacted with Walter back in the late 1940s, had made a farsighted statement back in 1962: "Does there exist some procedure for calculating the density that avoids altogether the use of 3N-dimensional space? Such a procedure might open the way to an enormous simplification of molecular calculations."[299]

Despite being a "pencil and paper" theoretician, Walter had a real ability for deriving general algorithms for use in quantum mechanical computational schemes. This was certainly true with the Kohn-Sham equations but also, to a certain extent, in his earlier research with Rostoker and with Luttinger. This was a difference between him and several other prominent solid-state theoreticians, including the Nobel Laureates Anderson, Mott and Van Vleck, whose contributions were more closely linked with explanations of specialised effects. In a certain sense, this aspect put Walter's main contributions closer in style to those of Slater.

In the 1970s and 1980s there were theoretical works which provided further foundation to the Hohenberg-Kohn theory of 1964. For example,

Levy published a more general proof and Lieb and Oxford showed there was a lower bound to the exchange-correlation energy.[300,301] There were then several further proposals for the exchange-correlation functionals that turned DFT into a theory that could provide quite accurate results for molecules. Expansions in terms of gradients of the electron density was a significant advance.[302] In a series of highly cited papers published in the 1980s and 1990s, Perdew and co-workers improved the forms further for the exchange-correlation energy.[303] Perdew stated:

> The exchange correlation energy is a relatively small part of the total-energy, but it dominates the binding energy and has been called "nature's glue". In a nutshell, the electrons swerve to avoid one another (like shoppers in a crowded mall) because of the Pauli exclusion principle and Coulomb repulsion of the electrons. This lowers the total energy, and does so more in bonded systems than in separated atoms because bonded systems have more nearby electrons to be avoided. This "dance of the electrons" is the largest part of what binds atoms together.[304]

A contribution that proved to be useful for molecules came from Lee, Yang and Parr in 1988 who fitted the exchange-correlation energies to electron densities and gradients obtained from accurate calculations of the correlation energy.[305] In addition, the Canadian theoretical chemist Axel Becke, also in 1988, included a fraction of the exact Hartree-Fock exchange energy in the exchange-correlation energy which allowed the exchange-energy density to have the correct asymptotic dependence.[306] The functionals for the exchange-correlation energy of both Becke, and Lee, Yang and Parr, were then combined into various forms (labelled by the initials of their names such as BLYP and B3LYP) which have been used very extensively.

In the words of Perdew: "It was Axel Becke who introduced empiricism into density functionals by fitting them to experimental results for bonded systems, but he tried to limit his empirical parameters to a few, and his empiricism was often only to refine a parameter that could already be estimated by theory alone."[304] Most importantly, these improved exchange-correlation functionals gave quite accurate results, with reasonable computational expense, for key quantities in chemistry such as optimised molecular structures, atomisation energies, chemical bond energies, vibrational spectra and other properties such as those needed to interpret nuclear magnetic resonance spectra.

The intensive improvements in DFT for chemical applications made in the 1980s came to a head at the 7th Congress of the International Academy of Quantum Molecular Science held in July 1991 in Menton in France. This academy was established in 1967 by a group of prominent quantum chemists including John Pople and Robert Parr. The famous Nobel Laureate Louis de Broglie was involved with the formation of the Academy. Every year the Academy elects a small number of new members who have advanced quantum molecular science. In addition, the Academy awards a medal to whom they consider to be the most deserving scientist in this field aged under 40. A major international Congress is also organised every three years.

In 1991 the President of the Academy was Robert Parr. Walter Kohn was one of the plenary speakers at the Congress and the title of his talk was "Density Functional Theory in Physics and Chemistry". He was elected into the Academy that year following the nomination of Parr:

> Walter Kohn is one of the world's most distinguished theoretical physicists, perhaps the most respected theoretical solid-state physicist active in the world today. He is of course the co-author of the founding papers on density functional theory in the 1960s with Hohenberg and Sham. He was, however, famous before that work and he has continued to produce important papers throughout his career, as for example illustrated in his recent works on the density functional theory of excited states. Notable is his care to be correct; he never makes extravagant claims. Not trained in chemistry, he nevertheless has attended and contributed much to a number of important chemical conferences.[307]

Becke was also a speaker at the Congress during which he was awarded the medal of the International Academy for that year (Figure 57). All of

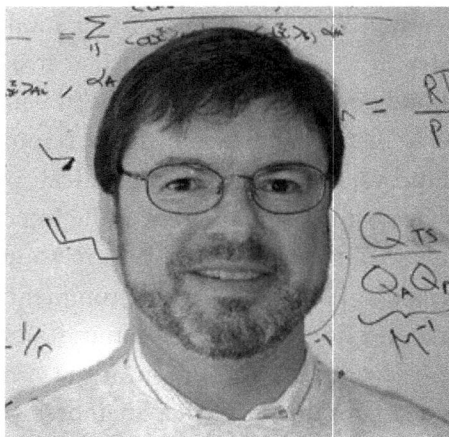

Fig. 57.   Axel Becke. (Courtesy of Axel Becke.)

these features helped to focus the meeting on DFT. The final talk was given by John Pople on "The Computation of Molecular Energies". With his co-workers, Pople had developed a general and widely used computer program for quantum chemistry calculations known as Gaussian. At the meeting Becke had a long five-course lunch over three hours with Pople in which DFT was discussed in detail. As Becke recalled:

> He [Pople] must have taken home a few things from our conversation, because within a year, he had published a paper on the B88 exchange energy density functional, and came out with a special release of the Gaussian program that included DFT. This had never happened before, and has not happened since. At that point, every chemist could do DFT calculations.[308]

Becke's talk at the meeting created quite a stir. He showed how his latest hybrid version of DFT could give results for quantities such as the energy required to break up molecules into atoms with at least the same accuracy, but at a fraction of the computational cost, of more conventional quantum chemistry methods (which were the subject of Pople's talk with his systematic

"G2" theory). I also gave a talk at the meeting in a different session. I recall my former Cambridge colleague and PhD supervisor, the quantum chemist Nicholas Handy, coming straight up to me after Becke's lecture and saying, in a very excited tone, "This is going to change quantum chemistry!"

Most scientists who go to a conference hope to return home with a few ideas from the meeting and make some new contacts. Very few conferences truly revolutionise a subject. That was certainly the case, however, with the 1991 International Academy of Quantum Molecular Science Congress in Menton. As soon as he returned home to Pittsburgh, Pople encouraged members of his research group, including Benny Johnson and Peter Gill, to implement the latest advances in DFT. At the Sanibel conference, held in Florida in the next summer of 1992, Pople was already reporting favourably on DFT which his group had implemented in the Gaussian code. In the next year, Pople and co-workers published a paper in the *Journal of Chemical Physics* that was to be highly cited and stated: "The density functional vibrational frequencies compare favorably with the ab initio results, while for atomization energies, two of the DFT methods give excellent agreement with experiment and are clearly superior to all other methods considered."[309]

There were similar conclusions from many other major research groups in quantum chemistry. For example, members of the Handy group in Cambridge had constructed a general quantum chemistry program called Cadpac. Handy's co-worker David Tozer wrote:

Nicholas Handy commenced his density functional theory (DFT) research in January 1992, following a recommendation from John Pople that he should investigate this alternative approach to the electronic structure problem. Handy described himself as being "demoralised by quantum chemistry" at that time, due to its excessive computational cost, and so he was immediately attracted to the simplicity of DFT with its potential for high quality results. Within a month, DFT had been implemented in Cadpac and a stream of publications followed.[310]

A similar example was given by Gustavo Scuseria from Rice University in Houston, Texas:

> In March of 1992, while the first author of this review [Scuseria] was actively working on coupled cluster theory, John Pople visited Rice University to deliver the Franklin Memorial Lecture. The subject of his talk was the impressive performance of the BLYP functional for thermochemistry, especially as judged by its low computational cost relative to the G2 theory. Pople was on his way to the Sanibel Symposium and gave the author a preprint which later became Pople's first publication on DFT. Pople credited the lecture by Axel Becke at the 7th International Congress of Quantum Chemistry in Menton in the Summer of 1991 as a turning point in his views on DFT. With the help of Peter Gill by email correspondence, the author promptly added a numerical quadrature code to his Hartree–Fock program and was comparing coupled-cluster results to DFT in a matter of weeks.[311]

One of the reasons these implementations could be carried out so quickly was that the Kohn-Sham equations had a similar structure to the Hartree-Fock form. Therefore, most of the computational procedures for implementing DFT were readily available, with relatively minor modifications, to the research groups who already had their own quantum chemistry computer packages. Following on from 1991, there was now an explosion of papers on DFT which has continued to the present day.[312]

Major challenges to DFT in 1991 were still the calculation of reliable activation energies for chemical reactions and the treatment of long-range dispersion forces between molecules. By further development of exchange-correlation functionals in hybrid and meta-hybrid forms, particularly through fitting to the results of accurate *ab initio* calculations for small molecules, and by improvements in the methods for dealing with long-range forces, these difficulties were then largely removed.[313] Further developments have also allowed for second- and third-row elements to be included within

DFT computations and even relativistic effects, important in heavy elements, can be treated.[314,315] Furthermore, through an ingenious algorithm developed by Roberto Car and Michele Parrinello, it has been possible to incorporate DFT directly into molecular dynamics calculations such as the simulation of liquids.[316] Another challenge was to extend DFT to excited states of atoms and molecules and this was achieved through a time-dependent theory developed by Erich Runge and Hardy Gross, who subsequently was a postdoctoral researcher and Heisenberg Fellow in Walter's group.[317]

Chapter Ten

Pople

Fig. 58. John Pople, 1961. (Courtesy of the Royal Society.)

John Pople was two years younger than Walter Kohn, being born on 31 October 1925 in Burnham-on-Sea, Somerset, UK (see Figure 58). In an interview, he described his upbringing:

> My family has no scientific background. As a child I lived in a small town in the West of England. My father was a shop keeper. He owned and ran the men's clothing store in this town. My mother's family were mostly farmers from different parts of England. So I had no professional scientific background in my family. As a child I did become extremely interested in mathematics at about the age of 11 or 12. I was by that time already keeping my father's accounts, large sums to be added

up in pounds, shillings and pence, which is quite complicated mathematics… When I started being exposed to algebra in school, I became extremely fascinated and spent much of my time teaching myself at the age of twelve more advanced parts of mathematics. In fact, I taught myself calculus using an old book which I found in a waste basket… and read it from cover to cover.[318]

Up to the age of ten, Pople attended local schools in Burnham-on-Sea. Then, in 1936, he attended Bristol Grammar School. He initially was a boarder at the school but, as was common with many boys at British schools, this was not to his liking so his parents allowed him to commute from his home in Somerset. This was two miles by bicycle, followed by 25 miles by train and one mile on foot. When the war started, the commuting became challenging particularly after the blitz-bombing of Bristol started in June 1940. Pople explained:

I continued to do this during the early part of the war, a challenging experience during the many air attacks on Bristol. Often, we had to wend our way past burning buildings and around unexploded bombs on the way to school in the morning. Many classes had to be held in damp concrete shelters under the playing fields. In spite of all these difficulties, the school staff coped well and I received a superb education.[319]

Pople enjoyed mathematics as he recalled:

I did not tell my teachers that I had taught myself calculus. And for a while I was reluctant for this to be known, and I used to make sometimes deliberate errors in mathematics classes so that I would not appear to be too smart. One feels the pressure from one's peers at that stage. But then after two or three years I sort of went public with this. Then the school authorities, this is at Bristol Grammar School which did have a very fine mathematics teacher, they told me that I should study for the scholarship exam for Cambridge. So I spent the last two or three years really preparing for this competitive intake at Trinity College Cambridge, which is where mathematicians tend to go in Britain, because they have a great history in development of mathematics.[318]

At Bristol Grammar School, Pople had a particularly good mathematics teacher Robert Lyness who had given his name to a periodic sequence called the "Lyness cycle". He also received excellent teaching in physics from T. A. Morris.[320] Pople took the Cambridge University Scholarship examination in 1942 during the middle of the war at the age of 17. At that time there was conscription into the armed forces of young men of his age. However, the British Government was already well aware of the importance of well-trained scientists and mathematicians for the war effort and this allowed Pople to go to Cambridge and take an undergraduate mathematics degree in just two years before going into government service.[320]

Trinity College, Cambridge, which was founded in 1546 by King Henry VIII, had a remarkable record in science and mathematics. Isaac Newton had made his great discoveries on classical mechanics, gravity, light and calculus at Trinity in the later part of the 17th century. Several of the most famous scientists had been Fellows of Trinity including James Clerk Maxwell, Lord Rayleigh, Ernest Rutherford, J. J. Thomson and William and Lawrence Bragg. By the time Pople attended Trinity, as many as 15 of its members had won the Nobel Prize. He had a formidable set of tutors. This included Cecil Powell who was to win the Nobel Prize for Physics in 1950 for the discovery of the pi-meson. In addition, there was the pure mathematician Abram Besicovitch who had been a referee for Fritz Rothberger, Walter Kohn's mathematics teacher in the Ripples Camp. Another tutor was Alan Wilson, who had studied with Heisenberg and had developed an early quantum mechanical theory explaining how energy bands determine if a material can be a semiconductor or a metal. Pople's studies went well and he achieved First Class Honours in mathematics.[320] Like Walter Kohn, Pople had been taught by excellent teachers in mathematics and physics at school and this had been followed by superb leading scholars at university.

Hoping he could do some research in fluid mechanics, Pople then went for a period in industry as he recalled:

> In 1945 as the war was ending, the government didn't quite know what to do with me because the war effort was winding down. So I was sent off into industry and I worked for the Bristol Aeroplane Company for two years, not doing anything very significant, until I could get back to Cambridge to start a research career. And so I returned in 1947. I spent one year doing sort of post graduate courses in all branches of applied mathematics and then it was after that that I decided to become a theoretical scientist. And at that point I picked on chemistry as a field that I would enter. But until then I had no background in chemistry. I'd given up chemistry at a quite early stage in high school.[318]

Back at Cambridge, Pople took the Part III Mathematics course for an extra year. His lecturers included Paul Dirac, Hermann Bondi, Nicholas Kemmer, Fred Hoyle and John Lennard-Jones.[320] So, like Walter Kohn, Pople did not have any formal qualifications in chemistry. He must have been stimulated by the lectures of Lennard-Jones as he chose to be supervised by him for his PhD starting in 1948. Lennard-Jones had been appointed as the first professor of theoretical chemistry at Cambridge in 1932. He had previously been a reader in theoretical physics at Bristol University. He was a pioneer on molecular orbital theory and intermolecular forces.[321] His first research student was Charles Coulson. The other members of staff in theoretical chemistry at Cambridge were Frank Boys and George Hall, whose research contributions would, in due course, influence Pople's career significantly.

Pople's first paper was published in the *Proceedings of the Royal Society A* in 1950 with Lennard-Jones and had the title "The molecular orbital theory of chemical valency. IV. The significance of equivalent orbitals."[322] The paper emphasised the significance of lone pairs of electrons in determining the shapes of molecules. Pople quickly followed up this work with a single-author paper in the same journal in which he described the electronic structure of the water molecule by two sets of equivalent orbitals pointing in near tetrahedral directions.[323] These papers were published using primitive calculations without the use of electronic computers. However, Pople's first papers were on molecular orbitals, the sophisticated computation of which would eventually win him the Nobel Prize. Pople extended his interest in lone pairs of electrons to the properties of liquids on which he published three papers under the general title "Molecular association of liquids".[324]

Pople obtained his PhD in 1951 and he was then also elected to a prestigious research fellowship at Trinity College which provided free meals and accommodation. He married his piano teacher Joy and drove a stylish MG sports car, showing something of a taste for the high life.[320] At the same time as Francis Crick and Jim Watson, he was elected to the elite Kapitza club in Cambridge where informal discussions were held with top international physicists. Previous visitors who had made dramatic announcements on quantum mechanics at the Kapitza Club had included Bohr, Heisenberg, Schrödinger, Landau and Born.[325] Pople continued to publish papers on statistical mechanics and intermolecular forces. In 1953, he also published a paper on "Electron interactions in an unsaturated hydrocarbon" that approximated electron-repulsion integrals by semiempirical parameters that led to the Pariser-Parr-Pople molecular orbital theory.[201]

Lennard-Jones left Cambridge in 1953 to become Principal of the new University College of North Staffordshire. Pople was appointed to a lectureship in mathematics. At this time, Robert Parr came from Carnegie Tech on sabbatical to Cambridge and their careers would then interact often (see Figure 59). Pople and Parr shared the vacated office of Lennard-Jones, and that was a good working environment to make friends (as Kohn and Hohenberg would also find).[326] Pople was just a little too young to be appointed to the vacant Chair of Theoretical Chemistry at Cambridge.

Fig. 59. The theoretical chemistry group at Cambridge University in 1953–1954. Back row (from left): Colin Reeves, Robert Nesbet, David Buckingham, Peter Schofield. Front row (from left): Mrs Scott (secretary), John Pople, Robert Parr, Frank Boys, George Hall, Victor Price, Alan Brickstock.

That went to Christopher Longuet-Higgins who was only two years older than Pople and was already quite famous for explaining, while he was an undergraduate at Oxford University, the bridged bonding by hydrogen atoms in boron hydrides.[327] Pople published an important paper in 1954 with Robert Nesbet (the postdoctoral supervisor of the current author) describing a computational method that became known as Unrestricted Hartree-Fock theory. This allowed molecular orbital calculations to be performed on molecular radicals.[328] Nuclear Magnetic Resonance was then becoming widely used in chemistry and Pople spent two sabbaticals at the National Research Council in Ottawa. He wrote a classic book with Schneider and Bernstein on *High Resolution Magnetic Resonance Spectroscopy.*[329]

However, by 1958 Pople was becoming disenchanted with the high teaching load and relatively low salary of a university lecturer in Cambridge, particularly as he and Joy now had three children. Also, as a lecturer in mathematics he preferred a post with a more applied scientific content.[320] The chance of a prestigious chair in theoretical chemistry in the UK in the next few years was low with Longuet-Higgins at Cambridge and Coulson at Oxford firmly in post.

Out of the blue in 1958, Pople was offered a permanent professorship in theoretical chemistry at the University of Florida, Gainesville with a ten-month salary of $13,000.[330] This university had major aspirations for a "Quantum Theory Project" and was soon to appoint the Swede Per-Olov Löwdin and also John Slater. This approach must have put ideas into Pople's mind of moving to the USA. However, at the same time he was offered the new position of Head of the Basic Physics Division of the National Physical Laboratory in Teddington near London and he accepted this post.[320] This was a civil service position with much administration and, although he made several impressive appointments particularly in the field of magnetic resonance, the next three years was a period of fairly limited research contributions from Pople himself. He missed key aspects of university life including a good library and research students.[319] It was not all negative for Pople at this time — he was elected a Fellow of the Royal Society in 1961 at the young age of 35.

We have already seen that scientific conferences played a major role in the research careers of Kohn and Pople. In the spring of 1961, Pople was involved with Coulson and Longuet-Higgins in organising a conference in Oxford on theoretical chemistry and Robert Parr was an invited speaker. During an interval, Parr, who recurs again and again in the careers of both Kohn and Pople, suggested to Pople that he take a sabbatical at Carnegie Tech for 1961–62 as Ford Visiting Professor. He jumped at the chance.[319]

Pople enjoyed his sabbatical year at Carnegie Tech despite the fact that Parr had, unexpectedly, moved on to John Hopkins University by the time he arrived in Pittsburgh. He had also just missed out on overlapping with Walter Kohn who had moved to La Jolla in the year before. Pople travelled widely in the USA with his family during the sabbatical year, which they much enjoyed, and he lectured in many university departments. With his broad research interests in chemical physics, Pople interacted well at Carnegie Tech and, with a vacancy for a theoretical chemist following the departure of Parr, it was put to him whether he would be interested in a permanent appointment there.[319] Pople wanted to be sure that Carnegie Tech was the right place for him and he asked Parr for his advice. Parr's response was surprisingly negative:

> I have taken quiet measures to assure that Illinois, Chicago, and Princeton think about whether they want you. Our country will profit much if we can entice you over here permanently, so I am anxious that you have the widest possible selection of opportunities.
>
> I think that you are much better able to evaluate the Carnegie Tech possibility than I am. I was so involved in an intensely personal way in trying to build up the school, and at times so frustrated in my attempts, that I probably no longer retain objectivity. I differed as to what the true potential of the Department is, and the best way to go about achieving the Department's goals... I felt that the administrative officers at Tech had outlived their usefulness. In my opinion they have not demonstrated in very recent years the courage to make the positive decisions that would have made the university move ahead at the speed it was capable of moving at.[331]

Thanks to Parr, the word had got round that Pople was interested in a post in the USA and he received several enticing offers. There was a battle to secure him from leading institutions which was somewhat similar to the efforts to secure Walter Kohn around 1957. Lee Allen, a theoretical chemist, made the case to his Department of Chemistry at Princeton:

> Stated simply, I believe that Pople is the best theoretical chemist in the world today. There is no single person in any area that we could presently attract who would produce such an impact on the standing of our Department. He is so widely known and enjoys such stature that I feel certain that he would

affect a discontinuity in the quality and number of applicants to our graduate program.

Pople has made central contributions to inorganic, organic, and physical chemistry (e.g., the leading electronic structure model among organic chemists today is no longer the Hückel scheme but the "Pople method"). He would help unify department research efforts and provide a new and stimulating professional spirit.[332]

Princeton made a tempting offer to Pople of $21,000 for nine months linked to Bell Labs who came in with $2,500 per month of summer salary.[333,334] Princeton also emphasised, in correspondence, the advantages of eminent colleagues such as Eugene Wigner (at the University) and Freeman Dyson (at the Institute for Advanced Study).[335] The University of Chicago, where Pople had been the Harkins Lecturer for 1962, was another university that came in strongly. Robert Mulliken wrote to emphasise the strength of theoretical chemistry colleagues there including Clemens Roothaan.[336] Their offer was a salary of $27,500.[337] The National Research Council of Canada, where Pople had spent two recent sabbaticals, also made an offer of $20,000 (Canadian) together with an appointment at Ottawa University.[338] However, in terms of salary, Carnegie Tech came out top.[339] Their offer was $23,000 for 9 months but with a vital top up of $7,000 from the nearby Mellon Institute in Pittsburgh where the physical chemist Paul Cross, a friend of Pople, was chairman.[340] Carnegie Tech offered funds for a research assistant and, most significantly, there would be no expectation that Pople would be required to do undergraduate teaching. His title of Carnegie Professor of Chemical Physics was essentially a research professorship.

Back in England the word had quickly got round that Pople was thinking of emigrating to the USA and there were several moves to keep him in the country. He was offered a Chair at Manchester University which had a fine record in physical chemistry.[341] Coulson at Oxford also wrote with a vague suggestion:

You are a former fellow of Trinity; and as such will have dining rights etc at Christ Church. But I am pretty sure that as soon as they knew that you were likely to be available, they would make you a Professorial Fellow. It's just the sort of thing that they would do. They're wealthy, and could do it without the worries that some of the smaller Colleges would have to face.[342]

As a previous President of a traditional Oxford college (Magdalen), the current author knows that this proposal would have been complicated to implement quickly, with a Statutory Chair being necessary at the University of Oxford for a senior appointment (a difficulty that also arose in Schrödinger's time in Oxford).[72] In addition, the Director of the National Physical Laboratory, Sir Gordon Sutherland, wrote to say that it may be possible to promote Pople to the position of Chief Scientific Officer which would give a salary increase and no administrative responsibilities.[343] However, Pople did not consider this to be competitive with the offers from the USA and his mind was made up to go to Carnegie Tech. By 1963, his salary in England was £5,412 which was equivalent to a figure very close to $15,000 with the exchange rate then prevalent.[344] Thus his offer from Carnegie Tech gave an irresistible increase in salary of a factor of two.

There was a furious row in Britain following the announcement of Pople's departure to the USA. It coincided with some other British scientists who were also leaving. There were numerous headlines early in 1964 in the popular newspapers including "Another one down the brain drain" and "Another scientist is off to America. My decision by top physics man."[345] The newspapers pictured Pople at home with his family. They stated that he was going for better research facilities and emphasised he would be doubling his salary in the USA. Compared to the annual salary of scientists, one newspaper stated "Pop singers get more in a week".[345]

The matter was even raised in the House of Parliament by Harold Wilson, who was then Leader of the Opposition and was very soon to be Prime Minister. He mentioned Pople and tabled a censure motion on the government for allowing leading scientists to leave Britain due to their poor facilities and salaries.[345] The newspapers reported that the issue was even mentioned by the Prime Minister Alec Douglas-Home in a meeting with President Lyndon B.

Johnson at the White House in Washington DC. However, LBJ seemed more concerned that Britain was selling buses to Cuba.[345]

At that time, Carnegie Tech did not rival Princeton nor Chicago for the reputation of its scientists but Pople had been at Cambridge University with some of the most eminent scholars such as Paul Dirac and this was not a main feature for him. He had spent a happy sabbatical with his family in Pittsburgh and was impressed with the local schools.[346] He always stated that his "home base" was very important to him.[319] So, despite the warnings from Parr, Pople decided for Carnegie Tech.

By 1962, electronic computers were becoming more widely available and it was becoming possible to calculate accurately and routinely the difficult integrals arising in molecular orbital theory. It was not now necessary to make the approximations of Pariser-Parr-Pople theory. Pople, therefore, had the foresight to consider a major new research programme in Pittsburgh in developing a general computer program for electronic structure calculations on molecules and he wanted to focus on this one topic. In this he was supremely successful.

There were visa issues that delayed Pople's arrival in Pittsburgh but he started there in March 1964. He gave a summary of his subsequent research over the next few years:

On my return to Pittsburgh, I resolved to go back to the fundamental problems of electronic structure that I had contemplated abstractly many years earlier. Prospects of really implementing model chemistries had improved because of the emerging development of high-speed computers. I was late in recognizing the role that computers would play in the field — I should not have been, for Frank Boys was continually urging the use of early machines back in Cambridge days. However, by 1964, it was clear that the development of an efficient computer code was one of the major tasks facing a practical theoretician and I learned the trade with enthusiasm. Mellon Institute, where I had an adjunct appointment, acquired a Control Data machine in 1966 and my group was able to make rapid progress... In 1967, Carnegie Tech and Mellon Institute merged to become Carnegie-Mellon University (CMU) and I remained on the faculty there until 1993.[319]

Pople mentions his former colleague Frank Boys. At Cambridge, Boys had shown that Gaussian functions formed a convenient basis set for the atomic orbitals of electronic structure of molecules because they simplified the complicated integrals arising over the electron repulsion terms.[347] Following on from this, Pople decided to call his new general purpose computer code for electronic structure calculations "Gaussian" and it was first released in 1970.[348] He, his research group and other colleagues implemented the major advances in *ab initio* molecular orbital calculations and continually modified and optimised this code as new developments were made. His research papers also showed a systematic test of the accuracy of the calculations against benchmark experimental data and continuing improvements of the models used. As we have seen in the previous chapter, Pople was supremely placed to implement the latest developments in DFT in his Gaussian code from 1991. Through this process, Pople fully established himself as a leader in the field of quantum chemistry. He had a realistic view of the mathematics of quantum mechanics: "I've spent my career trying not quite to solve these equations, but at least to get good enough approximation to them to become useful in actual chemical problems."[349]

In 1992, Pople was awarded the Wolf Prize for Chemistry.[350] This Prize is considered by many to be only second to the Nobel Prize. It was presented to him at a session of the Knesset in Jerusalem, Israel. The early work on the Gaussian computer package was done at Carnegie Mellon University and in 1993 Pople moved to Northwestern University, in Evanston near Chicago, as Trustees Professor. This placed him nearer the family of his daughter Hilary, and he lived in Wilmette, Illinois.

John Pople had a straightforward, business-like personality. He was a focussed, decisive and very successful person. He made his own luck. This is illustrated by a personal anecdote. I once accompanied Pople from Nice Airport to Menton on the Riviera in the South of France for an annual meeting of the International Academy of Quantum Molecular Science. Pople was driving and we passed the famous casino in Monte Carlo. He recounted the story that several years before he was similarly passing the casino with a French colleague Raymond Daudel. They agreed to go in the casino and each place 100 Francs on a number on the roulette wheel. If they won or lost they would then walk out. Pople said his number came up, he received his winnings of 3,700 Francs, he walked out and never returned.

Chapter Eleven

Santa Barbara

The University of California, Santa Barbara (UCSB) was established in 1944. It was the third undergraduate campus to be created in the University of California after Berkeley and UCLA. Like the La Jolla campus of the University of San Diego, it is based on a beautiful site next to the Pacific Ocean. In the 1970s, UCSB had a small but collaborative physics department with an emphasis on theory. An opportunity then arose for an exciting new Institute for Theoretical Physics (ITP), as explained by Jim Langer:

The idea for such an institute started with Boris Kayser at the National Science Foundation, who thought that our field needed some analog of a national laboratory. He solicited proposals, brought finalists to Washington to make presentations, and ultimately awarded the first (and, so far, only!) such grant to UCSB. The authors of the UCSB proposal were Jim Hartle, Ray Sawyer, Doug Scalapino, and Bob Sugar. Their plan was that the ITP would host about four different research programs per year, two in the fall and two in the spring, each focusing on a specific set of emerging research problems, and each involving ten or twenty visiting participants at any given time. To provide expertise and continuity for these activities, the ITP would have a director and a small group of permanent members, all of whom would have tenured appointments in the UCSB Physics Department. There would be postdoctoral associates in residence for two years or more. And there would be a broadly representative external advisory board to recommend programs and help recruit the senior scientists. The then incoming UCSB Chancellor, Bob Huttenback, played a key role by promising the resources needed to make this plan feasible.[351]

It was a surprise to the US physics community that the new institute was assigned to Santa Barbara as there were very strong bids from more famous institutions including Caltech and Yale. A key aspect of the UCSB proposal was that there would be a Director and two other permanent appointments. However, this needed clear institutional support. The new entrepreneurial Chancellor of UCSB, Bob Huttenback, who was subsequently involved with several controversies, was coming from Caltech. He had heard about the ITP initiative from Murray Gell-Mann who had told him that Caltech was sure to beat Santa Barbara in the competition. Huttenback responded to the four UCSB physicists making the bid: "So they want to play hardball. Tell them that I will guarantee the three positions and do everything I can to see that the Institute for Theoretical Physics becomes world renowned."[352]

A Director then had to be found and Jim Hartle, Ray Sawyer, Doug Scalapino, and Robert Sugar, who had written the original proposal for an ITP, selected Walter Kohn.[351] He had been a key player in founding and developing the new physics department at UCSD, and through his outstanding research papers and his warm character had become a highly respected member of the US theoretical physics community. In addition, solid-state physics research had become a major new area of theory and Walter was one of its leaders. He had not yet won the Nobel Prize but many observers thought (correctly) that this would happen in due course.

Walter also had a broad background and training in physics and had published in areas different to the solid state such as scattering theory and quantum electrodynamics. In addition, there might have been murmurings that Walter would be prepared to move the relatively short distance up the Californian coast from La Jolla for personal reasons, which are explained below. Furthermore, Walter had seen in operation the Institute for Theoretical Physics in Copenhagen run by Niels Bohr with its highly active programme of guest theoreticians. In addition, in his many visits to Paris, Walter had been involved in several international theoretical physics workshops run by the exuberant French-American Carl Moser through the organisation called CECAM. He liked the idea that the ITP would have workshops which would bring students and postdocs together with the leaders in a field for months at a time, rather than the days or weeks more common elsewhere. By 1979, Walter was aged 56 and the time was right for one last major move and a position of some responsibility. There were also other aspects in La Jolla that may have influenced his thinking to accept the Directorship of the new ITP in Santa Barbara.

Although his research was going well, the later part of the 1970s had been a complicated time in Walter's personal life. His wife Lois had been with him on his journey from Toronto, to Harvard, to Pittsburgh and then La Jolla. She had accompanied him on his many sabbaticals which she enjoyed. She found their two years in Copenhagen with a young family particularly special. She also had been with Walter on his numerous visits to France where the country and language were a lifelong interest for her.[92] In addition, when Walter went back to the University of Toronto in 1967 to receive his Honorary Degree, Lois and several members of her own family were proud to be present. By the late 1970s their three children, Marilyn, Ingrid and Rosalind, had grown up and were pursuing their education or careers. Walter was a very private man and rarely spoke or wrote publicly about Lois and his children. In 1977, Walter and Lois moved apart and divorced on 7 February 1978.[353] However, it should be emphasised that when Walter won his Nobel Prize in 1998 he stated in his Nobel Biography: "My first wife, Lois Kohn, gave me invaluable support during the early phases of my scientific career."[250]

It seems that there were also some problems in the Department of Physics at UCSD in the 1970s. Roger Revelle said:

> The one thing that did happen was that the physics department itself was a center of dissension. Not with the rest of the campus, but with each other. At least that's what Bernd [Matthias] told me when I talked to him about it one time. They couldn't ever

make up their minds about new appointments. Each was fighting for his area, or something. I'm not quite sure what. Several of them left. Bernd died. Walter Kohn got divorced and went to Santa Barbara. Marshall Rosenbluth wanted to work in a purely research environment, so he went to Princeton to the Forrestal Labs. He's probably, as I said, the best plasma physicist there is. Maria Mayer died. Carl Eckart died.[223]

Fig. 60. Mara Vishniac Kohn. (Courtesy of Laura Bialis.)

So when the call came for Walter to go to Santa Barbara there were also personal reasons that may have encouraged his move away from La Jolla and a fresh start. On 30 April 1978 Walter remarried. His second wife was Mara Vishniac (Figure 60). She was born in Berlin in 1926 and was the daughter of a well-known photographer Roman Vishniac. He was born in 1897 in St Petersburg in Russia but left for Berlin in 1918 due to anti-Semitic actions of the Bolsheviks. As anti-Semitism also took hold in Germany after Hitler's coming to power he was commissioned by an American Jewish Joint Distribution Committee, which was based in New York, to photograph the Jewish communities still based in Eastern Europe. He took over 16,000 photographs with the best equipment available and kept the negatives. He even entered internment camps in 1938 to picture the appalling living conditions.[354]

Following the Kristallnacht in 1938, Roman Vishniac's wife Luta took her daughter Mara and son Wolf to Sweden to get away from the Nazis. However, Roman was caught in France, where his parents were living, in the summer of 1940 and was interned by the Vichy authorities. The family had Latvian passports and this enabled them to obtain visas to go to the USA. They took a boat from Lisbon and arrived in New York on the last day of 1940. Roman Vishniac and his family managed to bring some of his original negatives of Jewish life to the USA and a number were brought by friends. However, only 2,000 of the original 16,000 were saved. They serve as a rare reminder of the Jewish communities in central Europe which were wiped out by the Nazis.[354]

To make a living in the USA, Roman Vishniac started a business in portrait photography and this included a celebrated photograph of Albert Einstein. Einstein was to declare this photograph his favourite and Walter would display a copy in his office in Santa Barbara. It is remarkable how many times Einstein reoccurs in a personal way in the Walter Kohn story: Walter's teacher in Vienna, Nohel, and his mentor in Toronto, Infeld, had both worked with Einstein, and Hertha Mendel's mother Toni was a close friend of Einstein in Berlin.

Mara Vishniac made it her duty to preserve and show the iconic photographs that her father had taken. Walter told the author of this book in 2001 that he considered it a major project for him also to support Mara in this work. She featured in a film about her father called *Vishniac* by Laura Bialis.[355] Mara and Walter had both experienced at first hand the brutality of the Hitler days and the Kristallnacht. This gave them a special bond. Mara had been married before and moving to Santa Barbara would be a new start for them both.

The ITP had a grant of $1m from the National Science Foundation for five years. The programme had to get going quickly and it had to be a success. The high-energy theorist, who was instrumental in the original proposal, Bob Sugar, was made Deputy Director. Walter also wanted to

Fig. 61. Walter Kohn, 1982. (Courtesy of American Institute of Physics, Emilio Segrè Visual Archives, Physics Today Collection.)

appoint a brilliant theorist in an area different to solid-state physics and his first major appointment was Frank Wilczek, a particle physicist who would win the Nobel Prize in 2004. Walter then soon appointed Jim Langer, who was another condensed matter theoretician whom he had taught as an undergraduate at Carnegie Tech and had subsequently spent a period on the faculty there.[351] A further condensed matter theorist to be appointed was the Nobel Prize winner Robert Schrieffer, the "S" of the BCS theory for superconductivity.

With his busy administrative duties, Walter's research productivity at Santa Barbara was low and he averaged just one paper a year in the five years when he was the Director of ITP (see Figure 61). There was a parallel with his start at UCSD some 20 years previously when he was quickly made the department chair. In the 1980s, with the new developments in DFT being applied to molecules, he was also in increasing demand from the international theoretical chemistry community to speak at their conferences. He had lost much of his hair and became distinctive for wearing a beret (see Figure 62).

Scalapino and Sugar recalled the early days of Walter's Directorship of the ITP:

Fig. 62. Walter Kohn at UCSB. (Courtesy of Department of Special Research Collections, UC Santa Barbara Library, University of California, Santa Barbara.)

In April 1978 the NSF Science Board approved funding for an initial period of five years, noting carefully that there was no certainty of continued support. So, in the fall of 1979 when Walter and his wife Mara arrived in Santa Barbara, the institute faced both doubts from the community and uncertainty of long-term NSF support. However, Walter's ability to relate to the outside physics community and his judgment regarding both physics and people soon began to change things. He persuaded leading members of the physics community to serve on the ITP Advisory Board and to put tremendous effort into the development of ITP's scientific programs. Under Walter's leadership, a wide range of programs, proposed by the community and by members of the Advisory Board, were scheduled and successfully run... Walter's creation of a welcoming atmosphere made the institute a place where physicists, ranging from new doctorates to senior leaders of the field, wanted to visit. Walter set the direction and established the atmosphere.[356]

Walter set the format which lasted for many years at the ITP. There were two five-month programmes in the first half of the year and two in the second half. Three programmes were usually in high-energy physics, condensed matter and astrophysics. The fourth was in a new direction. Two of these latter examples that Walter created were the topical, and still-running, questions: "How do we integrate Einstein's theory of gravitation with the rest of theoretical physics?" and "What are the implications of continued miniaturization in electronics?".[357] Walter also took the opportunity to make DFT a subject of a programme at ITP and the participants included Sham, Perdew and Levy.[304]

The Institute under Walter's leadership was a success, and places for the visiting scientists were much sought after by theoretical physicists from around the world. The ITP was instrumental in establishing UCSB as a major centre for science. Other institutes were set up elsewhere with a broadly similar format, with the Newton Institute at Cambridge University being one example. The National Science Foundation renewed the funding twice after enthusiastic reviews.[356] Walter handed over the Directorship to Robert Schrieffer in 1984 and five years later James Langer took over (see Figure 63). In due course, the ITP received a donation of $7.5m from the Kavli Foundation, which at that time was based in Santa Barbara. It was then renamed the Kavli Institute for Theoretical Physics (KITP) in 2001.

Fig. 63.   James Langer. (Courtesy of American Institute of Physics, Emilio Segrè Visual Archives, Gallery of Member Society Presidents.)

This Foundation was so pleased with the progress that it started to fund new Kavli Institutes around the world.

It should also be emphasised that Walter's warm personality was an important aspect in his success as Director of ITP (see Figure 64). There are many accounts of his support for scientists at all levels and his enthusiasm to hear about their research.[356] His background meant that his interests extended well beyond the solid state. He even had a broad knowledge on topics outside of science, including music, art (especially tapestries) and history. He was a very popular host, as Hardy Gross recalled:

> The parties on the Kohns' veranda are unforgettable. The marvelous view on downtown Santa Barbara and on the harbor, the beautiful garden (notably the loquat trees with their delicious fruits), Walter's mastership in telling stories and anecdotes and, last but not least, his talent of cooking made each of these parties a cheerful and memorable event.[358]

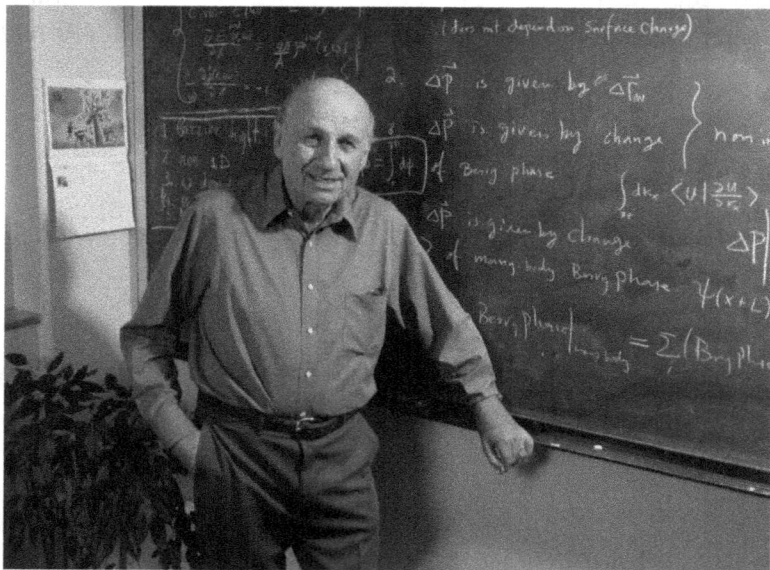

Fig. 64. Walter Kohn in his office in Santa Barbara, 1998. (Courtesy of Department of Special Research Collections, UC Santa Barbara Library, University of California, Santa Barbara.)

It is ironic that the one area where Walter had some criticism was in the computing facilities provided at the ITP. In the early 1980s, he was still very much a pencil and paper theoretical physicist who felt that computers could detract from the best research. He wanted to spend the funds raised for the institute on support for the visiting scientists as opposed to computer hardware. In 2004, Robert Sugar made a humorous comment on this aspect:

> Members [of the ITP] had to use computers located elsewhere on the Santa Barbara campus, which they accessed via dumb terminals located in a single room in the Institute. Walter insisted that terminals not be placed in the offices of Institute members for fear the overuse of computers would destroy their thought processes and distract them from doing real physics. Walter was, of course, an expert on this danger, since, as the inventor of density functional theory, he was then, and remains today, responsible for the burning of more computer cycles than any other person who has walked on the earth.[359]

Given everything he had been through as a young man, Walter's health had been good up until 1982. However, in January of that year, he had a five-vessel heart bypass operation by an expert surgeon Jack Love.[360] He recovered remarkably well after this and was very soon back in his Institute.

Walter was a familiar figure in rollerblading about the UCSB campus. For him, this activity brought back memories of ice skating in the cooler climate of the winters in Vienna, some 50 years before. He was something of a reckless athlete. Lu Sham recalls when Walter broke a shoulder skiing in 1965 and had to miss his first opportunity to talk on DFT at an American Physical Society meeting. Sham also recounts when Walter's boat sunk sailing off La Jolla and the coastguard was not amused.[361]

In 1988 Walter was awarded the President's National Medal of Science. Established in 1959 by the U.S. Congress, this medal is "the highest recognition the nation can bestow on scientists and engineers". The medal was presented to Walter at the White House by President Ronald Reagan on 15 July 1988. It was awarded "For his pioneering fundamental contributions

Fig. 65. Kohn Hall at UCSB, 2023. (Photograph by the author.)

to the theory of the electronic structure of solids, including the effective mass approach to defects in semiconductors, the so-called KKR method of band structure, and, most importantly, the density functional approach to the many-electron problem which has led to great advances in the understanding of bulk solids and solid surfaces."[362]

The ITP did not initially have its own building and operated in a central position on the UCSB campus on the top floor of Ellison Hall, with the lower floors housing History, Geography and Political Science. Quite soon the ITP had outgrown this space and a beautiful new building, which merges the Southern California and Spanish styles, was designed by the architect Michael Graves in 1993 on a spectacular site overlooking the Pacific Ocean (see Figure 65). It serves as a gateway to the southern end of the UCSB campus. It was called Kohn Hall in 1994 with the pre-Nobel dedication: "Named in honor of Walter Kohn, professor emeritus of physics, recipient of the National Medal of Science, who served as founding director of the Institute for Theoretical Physics from 1979 to 1984 and whose vision, leadership and international prominence enabled the Institute to achieve recognition as a pioneering center for advances in theoretical physics." A new wing was added in 2003 funded by the Kavli Foundation.

Walter continued in the Institute as Professor of Physics after he handed over the Directorship to Robert Schrieffer in 1984. He was made Emeritus and Research Professor in the important year for DFT of 1991 at the age of 68. In that year he was awarded the Eugene Feenberg Memorial Award in quantum many-body theory.[363] Scientists always are proud to receive a prize named in the detailed speciality of their research and this was certainly true for Walter in this case. In the spring of 1998 Walter was elected a Foreign Member of the Royal Society of London, a rare honour.[2] This was nearly 50 years after Schrödinger had been similarly elected, and Boltzmann had been elected 50 years before Schrödinger.[80] Walter was, therefore, in the company of the greatest physicists born in Austria.

Then, in the same year, he was awarded, with the Russian theoretical physicist Vitaly Ginzburg, the UNESCO Niels Bohr Gold Medal.[364] This is given "to recognise those who have made outstanding contributions to physics through research that has or could have a significant influence on the world." Given his close links with Niels Bohr, Walter often expressed his delight to win this medal with its broader implications beyond scientific research. However, there was very soon to be an even more exciting announcement.

Chapter Twelve

Nobel

The Nobel Prizes were created in 1896 through the will of Alfred Nobel in the subjects of Physics, Chemistry, Physiology or Medicine, Literature and Peace. The first three awards immediately became the leading prizes in science and have very much retained that prominence to the present day. Not only are the Nobel Prizes the most sought after in science, they have captured the imagination of the public around the world. There are few examples of real achievement that can beat the Nobel Prize.

The will of Alfred Nobel stated explicitly:

> The capital, converted to safe securities by my executors, is to constitute a fund, the interest on which is to be distributed annually as prizes to those who, during the preceding year, have conferred the greatest benefit to humankind. The interest is to be divided into five equal parts and distributed as follows: one part to the person who made the most important discovery or invention in the field of physics; one part to the person who made the most important chemical discovery or improvement... The prizes for physics and chemistry are to be awarded by the Swedish Academy of Sciences... It is my express wish that when awarding the prizes, no consideration be given to nationality, but that the prize be awarded to the worthiest person, whether or not they are Scandinavian.[365]

The Royal Swedish Academy of Sciences has always given special attention to the words of the will. For example, in physics "the most important discovery or invention in the field of physics" is stated while in chemistry the phrase has a slightly different emphasis on "the most important chemical discovery or improvement".

The Nobel Prize in Physics has a special kudos. This is because some of the physicists who have been awarded this Prize are as famous for their characters and life stories as their science. This includes, most prominently, Albert Einstein and Marie Curie. In addition, the pioneers for some of the great fundamental experimental discoveries such as X-rays by Röntgen (1901), radioactivity by Becquerel and the Curies (1903), the electron by J. J. Thomson (1906) and the neutron by Chadwick (1935) were all awarded the Physics Prize. Furthermore, fundamental new theory has to be proved by experiment and this, for example, was the case for the quantum theory of Planck (1918) and energy states of atoms by Bohr (1922).

There have been many controversies such as Einstein being award the prize for his prediction of the photoelectric effect and not for his theory of relativity, and also Ernest Rutherford who, for his iconic discoveries of the structure of the atom, was awarded the Prize for Chemistry and not for Physics. Quantum mechanics was invented in the mid-1920s but there were no awards for this breakthrough until there was clear experimental confirmation of the theory that came eventually through low-temperature heat capacity measurements and molecular spectroscopy.[72] Accordingly, Heisenberg won the Physics Prize for 1932, and Schrödinger and Dirac for 1933. Dirac was given the Prize for using quantum mechanics to predict a new particle, the positron, which was observed experimentally. Furthermore, there cannot be a Prize shared between more than three individuals and this has caused numerous challenges to the Nobel committees.

For each Nobel field, the Royal Swedish Academy of Sciences delegates a committee (typically of five or six members) to seek nominations and make a recommendation each year to the Academy. It is stated that: "The committee members are elected for a period of three years from among the members of the academy. In assessing the qualifications of the candidates, the committee is assisted by specially appointed expert advisers."[366] For the Prize in Chemistry, the committee sends out nomination forms each year to previous Laureates in Chemistry and Physics, members of the Royal Swedish Academy, professors in Scandinavian universities, and leading scientists in the field. At least one nomination is required to win the prize and, although a large number of nominations can make a Prize more likely, that is not guaranteed. For example, Arnold Sommerfeld received as many as 84 nominations for the Physics Prize over a period of 25 years but he was not successful.[367] In the case of Paul Dirac, just three nominations were sufficient, with one from the influential William Bragg.[72]

By 1998, a remarkable number of Walter Kohn's friends, co-workers or former colleagues who have been mentioned in this book had won the Physics Prize. This included Bohr (1922), Stern (1943), Bloch (1952), Shockley, Bardeen and Brattain (1956), Wigner and Goeppert Mayer (1963), Schwinger (1965), Bardeen, Cooper and Schrieffer (1972), Aage Bohr and Mottelson (1975), Anderson, Mott and Van Vleck (1977), Glashow (1979) and Bloembergen (1981). It must have rankled with Walter that he had not been added to this list.

There had not been very many Nobels in Physics for solid-state or condensed matter theory by 1998, as opposed to experimental breakthroughs. Lev Landau (1962, condensed matter theory), Bardeen, Cooper and Schrieffer (1972, superconductivity), Brian Josephson (1973, tunnelling supercurrents), Anderson, Mott and Van Vleck (1977, electronic structure of magnetic and disordered systems), Kenneth Wilson (1982, critical phenomena and phase transitions), and Pierre-Gilles de Gennes (1991, complex matter) are included amongst the number of such examples.

In experimental solid-state research there have been prominent examples going back to the diffraction of crystals by Max von Laue (1914) and William and Lawrence Bragg (1915). More recently there was, in addition to the semiconductor prize of 1956 for Shockley, Bardeen and Brattain from Bell Labs, Klaus von Klitzing (1985, quantised Hall effect), Gerd Binnig and Heinrich Rohrer (1986, scanning tunnelling microscope), and Georg Bednorz and Alex Müller (1987, superconductivity).[368] Rather surprisingly, Felix Bloch, who introduced his very widely used Bloch waves for electrons in periodic lattices in 1928, did not win the prize for this early theoretical advance but he did win the prize in 1952 for his work on Nuclear Magnetic Resonance.

The details of the nominations for the Nobel Prizes are kept secret for 50 years. John Perdew and Bengt Lundqvist both have stated that they nominated Walter for the Physics Prize and it is likely that there would have been several other such nominations.[369,370] However, the basic idea of using electron densities as opposed to wavefunctions, as formulated in the Hohenberg-Kohn paper, had already been considered by Thomas, Fermi and Dirac in the very early days of quantum mechanics and Slater had developed his own variant of the approach. So the basic claim to the originality of the idea may have been challenged. The view of Anderson would have been influential as he was a Nobel Laureate who was still active in solid-state theory — his co-Nobel Laureates and close friends of Walter, Van Vleck and Mott, had died in 1980 and 1996, respectively. On Walter's research Anderson wrote: "It is Walter's great strength to find important new physics in areas which ordinary people consider to have been long since mined out."[203] This comment perhaps links back to Thomas-Fermi-Dirac. There was also the significant computational advance in the Kohn-Sham paper. However, much of the research citing that paper was coming from chemistry departments. So it may have been hard to make an overwhelming case for Walter to the physics committee, but things were different in chemistry.

By 1998, there had been few Nobel Prizes for pure theory contributions to chemistry. Linus Pauling won the Chemistry Prize in 1954, mainly for his application of valence bond theory, Robert Mulliken in 1966 for molecular orbital theory and Roald Hoffmann and Kenichi Fukui became Nobel Laureates in 1981 for their orbital theories of chemical reactions. In addition, Rudolph Marcus won the Chemistry Prize in 1992 for applying wave mechanics to electron transfer reactions in chemical systems. There have also been some prizes in which quantum theory was an important component such as the prize for Gerhard Herzberg in 1971 for spectroscopy, that to William Lipscomb in 1976 for boron chemistry, and the 1986 Prize to Dudley Herschbach, Yuan T. Lee and John Polanyi for research in molecular beams and chemical dynamics.[72] However, until 1998 there had been no prize for computational chemistry.

The International Academy of Quantum Molecular Science held the key Congress in 1991 that communicated the progress in making DFT useful in chemistry and initiated the huge rise in DFT applications and citations.[312] That Academy is made up of elected individuals who have contributed to quantum molecular science and has included many notables already mentioned in this book including de Broglie, Pauling, Fock, Van Vleck, Slater, Mulliken, Roothaan, Teller, Herzberg, Lipscomb, Hoffmann, Fukui, Marcus, Karplus, Parr, Pople and Kohn.[371] From 1991 to 1997 the President of the Academy was Robert Parr. In the summer of 1997, John Pople took over the Presidency and the Treasurer was Björn Roos. The members of the Academy have frequently been invited to nominate for the Nobel Prize in Chemistry and this has put the Academy in an influential position.

Robert Parr, in particular, was a very strong supporter of Walter Kohn and nominated him for several major prizes related to chemistry during the 1990s. For example, in January 1998 he nominated Walter for the $500,000 Welch Award in Chemistry. This is the leading prize for chemistry in the USA. This Prize has the purpose "to foster and encourage basic chemical research and to recognize, in a substantial manner, the value of chemical research contributions for the benefit of humankind."[372] This Award was initiated in 1972 and nearly all the winners up to 2022 have been experimental chemists with the exception of E. Bright Wilson, Kenneth Pitzer, William H. Miller and Noel Hush. In his nomination of Walter Kohn for the Welch Award, Parr wrote:

Since WWII, quantum chemistry slowly has developed into a quantitative science, actively used by thousands of chemists and biochemists around the world… While the development of quantum chemistry has been slow, it sped up considerably when DFT methods began to get into quantum chemists' computer programs about ten years ago. There has been a sudden acceleration of the applications of quantum chemistry to larger molecules, and even to biomolecules. The person most responsible, and singularly so, is Walter Kohn, because he founded DFT.

DFT is not just a different scheme for solving Schrodinger's equation. For ground states it is the equivalent of the Schrodinger equation, yes, but it is cast in a fundamentally very different and extraordinarily useful and physically transparent form. Namely, the electron density is the basic entity in the theory, not the wavefunction. This makes all the difference, because the basic equations are now equations in functions of three variables (x,y,z), not in the space and spin variables for all N electrons (4N variables).

When the delicate questions are resolved as to how to approximate the so-called exchange-correlation potential in the theory, DFT calculations become feasible for systems with thousands of electrons. Various DFT techniques now are the method of choice for medium to large molecules, and are being employed extensively by just about everyone...

Kohn invented DFT in two magnificent papers in 1964 and 1965 with Hohenberg and Sham, respectively... Density functional theory is of the utmost importance for the understanding of the electronic structure of chemical species, an importance comparable to the importance of the greatest classic works in the subject. As the inventor of DFT, Walter Kohn would be a wonderful choice for the 1998 Welch Award.[373]

Parr's enthusiastic nomination of Walter for the Welch Award was not successful and it went in 1998 to the French biochemist Pierre Chambon. Every Welch Award has been given to someone working in a chemistry or biochemistry department and perhaps the nomination of a theoretical physicist was too far in 1998. It is quite possible that Parr used the same phrases to nominate Walter for the Nobel Prize in Chemistry. However, the details of the nominations will not be released until 2048.

Through the 1990s, John Pople was thought by many to be the most respected living computational quantum chemist and he had won the prestigious Wolf Prize in 1992. His Gaussian computer package had become the most widely used for quantum mechanical computations on molecules. He had not been the originator of most of the algorithms used in the Gaussian code but he and his collaborators had brought the techniques together, systematically improved and benchmarked them, and made them computationally efficient and straightforward to use. Thus many scientists were using Gaussian to calculate key chemical quantities such as molecular structures, bond energies, spectra and reaction rates.

In 1998, there were also several other powerful quantum chemistry computer packages that were being used extensively around the world. These included Columbus (Isaiah Shavitt, USA), Cadpac (Nicholas Handy and Roger

Amos, UK), Molpro (Hans-Joachim Werner, Germany and Peter Knowles, UK), Aces (Rodney Bartlett, USA), Berkeley (Fritz Schaefer, USA), Jaguar (Richard Friesner and William Goddard, USA), Gamess (Martyn Guest, UK and Mark Gordon, USA), Molcas (Björn Roos, Sweden) and Turbomole (Reinhart Ahlrichs, Germany). Here, leading scientists involved with the software package are named but there were numerous other co-workers as well. Indeed, a feature of Gaussian was that many quantum chemists contributed new algorithms to keep the code up to date and make it more efficient and powerful.

By 1998, it was thought widely by the community of quantum chemists that John Pople would receive the Nobel Prize for Chemistry. It was sometimes said, with humour, that Pople was the Pope and the other leading quantum chemists were the cardinals. I know that my influential colleagues at Cambridge University, David Buckingham and Nicholas Handy, were nominating Pople throughout the 1990s. However, there were many quantum chemists who had made useful contributions to the field and Pople did not have a standout theory that he had truly invented himself in *ab initio* quantum chemistry. The opposite was true with Walter Kohn. He had invented such a theory but had not developed it and applied it in chemistry. However, the Kohn-Pople combination ticked the boxes, especially as Pople had played such a key role in popularising DFT after 1991.

Another key factor was Björn Roos. He was not only the Treasurer of the International Academy of Quantum Molecular Science but was also, most crucially, on the five-member Nobel Committee for Chemistry that made the recommendation to the Royal Swedish Academy of Sciences. A highly respected quantum chemist, he was very well informed on the subject. The other full members of the Nobel Chemistry Committee for 1998 were the biophysical chemist Bengt Nordén, the structural chemist Torvard Laurent, the biochemist Carl-Ivar Brändén, and the organic chemist Lennart Eberson, who was the chair.

The question has sometimes been raised as to why Hohenberg and Sham did not win the Prize with Kohn. The Nobel Prize has a maximum of three winners in each year and so Pople would have had to be omitted if they were included and it is unlikely they would have been nominated for the Chemistry Prize. From the statements made after the Nobel announcement, it is clear that the award was made also for Pople's contributions to other areas of quantum chemistry, not just to DFT. It is probably true to say, however, that Pople would not have won the Prize without Kohn, and Kohn would not have won it without Pople. Furthermore, a significant catalyst for them both was Robert Parr who had been a colleague of Kohn at Carnegie Tech in the 1950s, had got to know Pople on his sabbatical to Cambridge in 1953–4, had been a key player in Pople's move to Carnegie Tech in 1964, and had been the first quantum chemist to pick up DFT in a major way.

To add a minor personal anecdote. In the summer of 1997 I co-organised a conference for the retirement of David Buckingham from his Chair of Theoretical Chemistry at Cambridge.[374] The plenary speakers at the meeting included John Pople (the PhD supervisor of Buckingham), Ahmed Zewail (then the co-editor with Buckingham of the journal *Chemical Physics Letters*, which had published several of Pople's most highly cited papers) and Björn Roos. After the meeting, my wife Heather gave a lift to the railway station and in the back of the car sat Pople, Roos and Zewail. Pople was to win the Nobel Prize a year later, and Zewail the year after that.

When the decision of the Royal Swedish Academy of Sciences was made on the morning of 13 October 1998, Björn Roos attempted to telephone John Pople at home in Wilmette, Illinois but he was out of town at the time. A message was left on his answer phone.[346] Roos then made the famous call to Walter Kohn, which is described at the start of this book, when he explained that the Nobel Prize was to be in Chemistry, not Physics (see Figure 66 of

Fig. 66.   Mara and Walter Kohn, just after they had received the Nobel Prize telephone call on 13 October 1998. (Courtesy Getty Images.)

jubilant Walter and Mara after the announcement).[1] In the statement released by the Royal Swedish Academy of Sciences, it was announced that the prize was awarded "to Walter Kohn for his development of the density-functional theory and to John Pople for his development of computational methods in quantum chemistry". A longer press release contained the statements:

Researchers have long sought methods for understanding how bonds between the atoms in molecules function. With such methods it would be possible to calculate the properties of molecules and the interplay between them. The growth of quantum mechanics in physics at the beginning of the 1900s opened new possibilities, but applications within chemistry were long in coming. It was not practically possible to handle the complicated mathematical relations of quantum mechanics for such complex systems as molecules...

Things began to move at the beginning of the 1960s when computers came into use for solving these equations and quantum chemistry (the application of quantum mechanics to chemical problems) emerged as a new branch of chemistry. As we approach the end of the 1990s we are seeing the result of an enormous theoretical and computational development, and the consequences are revolutionising the whole of chemistry. Walter Kohn and John Pople are the two most prominent figures in this process. W. Kohn's theoretical work has formed the basis for simplifying the mathematics in descriptions of the bonding of atoms, a prerequisite for many of today's calculations. J. Pople developed the entire quantum-chemical methodology now used in various branches of chemistry...

Walter Kohn showed that it is not necessary to consider the motion of each individual electron: it suffices to know the average number of electrons located at any one point in space. This has led to a computationally simpler method, the density-functional theory. The simplicity of the method makes it possible to study very large molecules. Today, for example, calculations can be used to explain how enzymatic reactions occur...

John Pople is rewarded for developing computational methods making possible the theoretical study of molecules, their properties and how they act together in chemical reactions. These methods are based on the fundamental laws of quantum mechanics as defined by, among others, the physicist E. Schrödinger... Pople made his computational techniques easily accessible to researchers by designing the Gaussian computer program. The first version was published in 1970. The program has since been developed and is now used by thousands of chemists in universities and commercial companies the world over.[375]

John Pople had just been awarded the Prize for Theoretical Chemistry of the American Chemical Society shortly before the Nobel announcement. He was visiting his son Mark on 13 October in Houston, Texas where he was having breakfast in a hotel with his wife Joy. He then had a call from his daughter Hilary with the exciting news of the Nobel Prize. She had heard it from a friend who had seen the announcement on the internet. Almost at the same time Pople saw the announcement on the TV news.[346,376] He put out a press release with the statement: "I consider this to be a great honour not just for myself but for all the students who have worked with me over the years."[377] He would repeat this type of phrase several times in the coming weeks and months — the Gaussian code was very much a team effort.

Even in 1998 there was the possibility of a false announcement or telephone call and the Laureates were pleased to receive almost identical faxes on 13 October from Erling Norrby, the Secretary General of the Royal Swedish Academy of Sciences, confirming the awards. For Walter, this stated:

Dear Professor Kohn,

Hereby I have the honour and pleasure to confirm in writing that the Royal Swedish Academy of Sciences has decided to award the 1998 Nobel Prize in Chemistry with one half to you for your development of the density-functional theory and one half to Professor John Pople, Northwestern University for his development of computational methods for use in quantum chemistry.

On behalf of the Academy I congratulate you warmly and look forward to meeting you when you come to Stockholm. You will receive your Prize from the hands of His Majesty the King on December 10th. A number of arrangements will precede and follow this event.

According to the statutes of the Nobel Foundation, each Laureate is required to give a lecture within six months after the day the award was presented. The subject of the lecture should be on the work for which the Prize was made and will be published in "Les Prix Nobel". In accordance with the tradition Laureates give their lectures during the visit in Stockholm in December...

On December 7th at 5.00 p.m., the Academy gives a reception in honour of you and the other Laureates in Physics, Chemistry and Economic Sciences and their families. I am also pleased to invite you and your wife to a dinner following the reception at the Academy to meet the other Laureates and the Members of the Nobel Committees and their wives.[378]

The arrangements were essentially the same format as for the previous 97 years of the Nobel Prizes — Erwin Schrödinger received a very similar letter when he was in Oxford in 1933.[72] A famous aspect of the Nobel Prize is that when the winners are first informed of their award they are told they have 30 minutes to comprehend the implications and then their lives will change forever. To a certain extent, that was true with Walter Kohn. He was the first member of the faculty of UCSB to win a Nobel Prize, but there would soon be several more in subsequent years.

During 13 October 1998 and over the next few days Walter received hundreds of congratulatory email messages (then the preferred method of communication of most scientists), a few faxes, many telephone calls and media requests of interviews from all around the world. As was also the case with the previous Austrian Nobel Laureate for wave mechanics, Schrödinger, Walter carefully filed all the congratulations, and these remain in the Special Collections at the UCSB.[379] Below we describe some of these messages.

Walter's family quickly emailed their delight at the award. His sister Minna said: "Thanks for the invitation. Please Vienna – Stockholm – Vienna. I'm completely touched. With Love Minna." Walter's daughter Marilyn had already congratulated her father on the phone just after he had heard the news. She then also emailed: "Daddy, … let me know if and when you will be in the Bay Area (any time in the next year or so), as I have a lot of friends who want to meet you."

One of the first actions that Walter took after hearing the news was to email Hohenberg, Sham and Parr collectively inviting them to the celebrations in Stockholm:

> I would be enormously pleased if you would be my guests at the celebrations in Stockholm, Dec 7–10, inclusive. This would give me an appropriate opportunity to thank you for your major and very different roles leading up to the formal recognition of DFT as an important contribution to science... It will be my pleasure to take care of your travel and hotel expenses.

Pierre Hohenberg had already emailed simply: "What great news! My heartfelt congratulations." Lu Sham wrote to say that unfortunately he could not come to Stockholm due to health reasons. He also wrote:

> First, let me congratulate you again for the Nobel Prize. It is a well-deserved recognition for all the lasting contributions you have made to condensed matter physics, especially the

founding of the density functional theory. It is also wonderful that the recognition has come from the chemists, who, in the early day, were our severest critics... Actually, speaking of thanks, I want to take this opportunity to thank you for the wonderful postdoc training which started my career and for getting me back to La Jolla.

It is possible that both Hohenberg and Sham had been considered to share the Physics Nobel Prize with Walter but they never showed any sign of rancour. They always looked up to him as the leader in the development of DFT and, after his death in 2016, Hohenberg with Langer wrote Walter Kohn's Biographical Memoirs for the Royal Society and the National Academy of Sciences.[2,380] In addition, the view has sometimes been expressed that Parr could have been the third name on the Chemistry Prize as he played a unique role with both Kohn and Pople. However, he always spoke very favourably about them.

Some of the congratulations had a particular pleasure for Walter. For example, from the University of Toronto Steve Halperin, and Pekka Sinervo, Chairs of Mathematics and Physics (respectively), wrote:

Given your roots here at the University of Toronto, we in the Departments of Mathematics and Physics take great pride in your achievements, not the least being this most recent honour. The Mathematics and Physics programme remains one of this University's "crown jewels" and as an alumnus of it your achievements add to its lustre. We hope that we will be able to host you at the University some time in the future.

Walter replied with some humour and also an indication of the very busy schedule he was accumulating:

I strongly believe that the University of Toronto offered me the best education in mathematics and physics available anywhere in the world in the 1940's, when I was a BA and MA student there. Individual professors whom I remember particularly vividly include Beatty, Pounder, Infeld, Coxeter, Brauer, Barnes, Hogg, Crawford, Burke, Weinstein. I have been deeply grateful ever since.

In light of my unexpected Nobel Prize in CHEMISTRY I find it amusing to remember that, in spite of Dean Beatty's strong intervention and my own feeble efforts, the Chair of Chemistry, Professor Kenrick, prohibited me and a few others from entering the Chemistry Building, because of our Austrian/German nationality, and thus excluded us from all college level Chemistry courses. 2 [and a] 1/2 years later, still an "enemy national", I was accepted by the Canadian Infantry Corps as an oversees volunteer, while the Mounted Police continued to keep a close eye on me. One year later, in 1945, I received my BA "on active service" (having missed my senior year), and the following year, now a Canadian national, my MA in Applied Mathematics. How is this for "Theatre of the Absurd"? I would be very happy to visit Toronto for a Welsh or some other lecture. My 1999 schedule is already enormously full. Could we think of the year 2000?

His friendly rival Anderson emailed referring to Rutherford's Nobel in Chemistry and not Physics: "THE BIG ONE. If it's good enough for Rutherford, it's good enough for you. A wonderful choice, a wonderful idea — just right!!" Walter's former co-worker and Nobel Laureate Nico Bloembergen wrote: "Longevity appears to be a necessary, although not a sufficient, condition to receive this honor." Bob Schrieffer, also a Nobelist and a former Director of ITP, emailed: "Your contributions have made a great impact in physics and chemistry."

Ed Creutz had been a key mentor, supporter and provider for Walter over many years, starting at Carnegie Tech in 1950. He wrote with humour:

Fig. 67. Walter Kohn learning how to do chemistry experiments. (Courtesy of Department of Special Research Collections, UC Santa Barbara Library, University of California, Santa Barbara.)

"I am delighted to know you have won THE PRIZE. A long way you have come from gravitational experiments with a weighted rubber sheet!"

There were a huge number of congratulations from former students, colleagues from UCSB and friends in Santa Barbara, where Walter was very popular in the community. With so many messages Walter had a humorous card produced which showed him learning how to do chemistry experiments, something he had, of course, never actually done (see Figure 67). He would send this card in replies to his friends.[379]

Walter was always delighted to hear from those who were in the Ripples Camp in New Brunswick with him during the war. Ernest Eliel wrote:

Congratulations on winning the Nobel Prize (even if it was in chemistry rather than physics, but we are broadminded and all appreciate how much your work has benefitted chemistry)! I guess if we had had to bet on someone in Camp T to win the prize, it would have been you! (And I say this even though we were sometimes at odds then!).

Eliel had escaped to Scotland from Berlin in 1938 and had been on the *Sobieski* with Walter from Glasgow to Canada. There they had a somewhat bitter academic row over whether physics was more important than chemistry, which was very ironic with Walter eventually winning the Nobel Prize for Chemistry. After one year in the Ripples Camp, Eliel's family managed to arrange a visa for him to go to Havana, Cuba where he studied chemistry as an undergraduate. He then went to the University of Illinois for his PhD and became one of the leading chemists in the USA, being elected to the National Academy of Sciences, writing an important book on the stereochemistry of organic molecules, being made President of the American Chemical Society and winning its leading award the Priestley Medal. He became a professor of chemistry at the University of North Carolina where he was a close colleague of Robert Parr. Like Walter he became active in several causes and won awards for promoting science in South America.[381]

Beat Jost, son of Res Jost, wrote a poignant message: "Yesterday I heard it on the news that you received the Nobel Prize for Chemistry. You can't imagine how happy I was. I'll never forget that you were with Res when he died. I'm convinced that he would be just as happy for you." Philippe Nozières commented: "I feel particularly excited by the citation. As you may remember the density functional approach was elaborated in my office at École Normale, when you were on sabbatical! I did not foresee then that you would become a chemist." Nicolas Rivier from the Université Louis Pasteur said: "I hope that your uncle is still alive and was able to phone you at UCSB to congratulate you in French."

Roger Newton had written the classic text *Scattering Theory of Waves and Particles*.[382] He said: "When I think back now on my graduate student days, all but one (Wendell Furry) of the physicists from whom I took courses won the Nobel: Julian [Schwinger], Van [Vleck], Bridgman, Norman Ramsey, and now you. But chemistry? Of course, you are in good company with Rutherford."

In the same year as Walter's prize the Physics award had gone to Robert Laughlin (a theoretician from Stanford), Horst Störmer (Columbia) and Daniel Tsui (Princeton) for their discovery of the fractional quantum Hall effect — another prize for condensed matter physics. Douglas Osheroff had overlapped with Walter at Bell Labs and had then moved to Stanford. He had won the Physics Nobel in 1996 for discovering superfluidity in helium-3. Steven Chu had won the Prize in 1997 and this meant Stanford had a Physics Laureate three years in a row. Osheroff wrote to Walter: "Only a few Nobel Prizes go to really great human beings, and you are definitely one of them … Needless to say, things were a madhouse here yesterday. You'd think that after so many times it would be ho-hum, but it never is. Bob [Laughlin] is grinning ear to ear." He also gave advice to Walter on pacing himself with all the lecture and media invitations, and to bring ten guests and pay for them from the prize money before any taxes are taken: "My advice is, bring as many friends as you can. It is a great time, they will remember it always, and you will never ever regret having spent the money!"

Several other solid-state physicists had written to Walter to say how pleased they were that the Nobels had gone to their subject in both physics and chemistry in the same year. Anthony Leggett from the University of Illinois, who was to win the Physics Prize in 2003 for theoretical work on superfluidity, wrote: "My only slight regret is that they apparently could not figure out a way to award it jointly in physics and chemistry." David Mermin, who had written the paper to show that the Hohenberg-Kohn theorems can be extended to finite temperatures, said he was delighted with the news and added: "I still think you have done more for physics than you have for chemistry." Jorge Hirsch from Walter's old department at UCSD, and famous for inventing the widely used h-index for scientific publications, wrote: "Sometimes chemists can be smarter than physicists!" Lillian Hoddeson from the University of Illinois, who has written extensively on the history of modern physics including the solid state, said: "I'm sorry Quin could not be around to enjoy this extreme honor with you."

Walter's surgeon Jack Love, who had carried out Walter's heart bypass operation in 1982, remarked with humour: "Since the prize is not given posthumously, I feel I may have had a hand in keeping you among us for this momentous occasion (and, I trust, for many more years to come)."

The Swedish theoretical chemist Per-Olov Löwdin had done important work on density matrices used in DFT. He wrote: "Recalling the fact that G.N. Lewis, Arnold Sommerfeld, Walter Heitler and Fritz London, Friedrich Hund, Erich Hückel, Henry Eyring and John C. Slater and several others never got the Prize, one realizes how difficult it is for a theoretician to get this award, and what a tremendous impact John Pople and Walter Kohn must have had on the ordinary chemists." Bengt Lundqvist, an influential solid-state theorist from Chalmers University in Sweden, said: "I have been proposing this for a decade now, and it is really great that the significance of your work has resulted in this reward." Another influential Swede, the materials scientist Börje Johansson, said that when he had met Walter in Crete in the summer of 1998 he knew how close he was to the Prize in Chemistry, but, as a member of the Royal Swedish Academy of Sciences, he could not tell him. The physicist Robert O. Jones, who had made many developments and applications of DFT, emailed: "I would never have thought it possible that the chemists would have done it." John Nohel, a mathematician who had moved from the University of Wisconsin to Zurich, wrote: "Emil Nohel would have been pleased." Walter was also particularly delighted to receive the congratulations of Emil Nohel's son who was living in Israel.

Walter also had a message from David Tambo who was Head of Special Collections at UCSB. He said that they would be very pleased to have Walter's collected papers. Walter received a similar request from UCSD. The papers did, however, eventually go to UCSB in over 100 boxes containing most of the letters and documents associated with Walter from 1950 onwards.

Walter was contacted by TV and media from all over the world to give interviews. CNN was quick off the mark on the morning of 13 October emphasising that the research was done at American universities. The *New York Times* report had the heading: "Five Scientists from the United States Win the Nobel in Chemistry and Physics Today for Work Exploring the Inner Structure of Matter." Walter was asked about the Chemistry, and not the Physics, prize and replied: "You may think they made a mistake… Science has a kind of unity, interdisciplinary acts can be fruitful."[377] It was also mentioned that the Prize in Physiology or Medicine had gone to three Americans Robert Furchgott, Ferid Murad and Louis Ignarro for their research on the role of nitric oxide in biology.

The media from Austria were particularly persistent for an interview but Walter was resistant to their demands as he and his family had been so badly treated there in the late 1930s. He always felt that Austrians were guilty for allowing the Nazis to take over so easily after the Anschluss. He did agree to a recorded interview with the Nobel Foundation and also with some local outlets. He was also invited to lecture in the main universities in Sweden, Finland, Denmark and Norway. He was pleased particularly to receive the congratulations and invitation from his friends and previous Nobelists Aage Bohr and Ben Mottelson from the Niels Bohr Institute in Copenhagen. There had been a tradition of over 70 years for the winners of the Physics Prize to lecture there just before or after the presentation in Stockholm.

On the afternoon of 13 October there was a press conference very appropriately held in Kohn Hall on the Santa Barbara Campus. Walter was pleased to see there the "Gang of Four" Jim Hartle, Ray Sawyer, Doug Scalapino, and Bob Sugar who had recruited him to the ITP. Even though he had been woken at 5 am by a most exciting phone call he was able, at the press conference, to give a detailed and lucid summary of his life and career. He was introduced by David Gross, who was then ITP Director and would win the Nobel Prize for Physics in 2004. The Chancellor of UCSB Henry Yang was ecstatic about the first Nobel for his University and stated: "Walter, your inspiration and impact on the proud UCSB campus is beyond what I can describe today." Yang said he wanted to arrange a black-tie dinner for Walter, but Walter was never a fan of the black-tie and subsequently wrote to Yang to say so. Doug Scalapino also told those present:

> Walter's theoretical ideas of how you compute the properties of electrons in molecules, this applied to molecules, it applies to chemical reactions, it is very important in biology, it is very important in polymers. All of this is true. Those of us in physics, though, know how important it is in determining the electronic band-structure of matter, in determining the properties of semiconductors, in determining the properties of superconductors, in determining, basically, the properties of matter. This is the reach that I think is so remarkable and that I think is what we celebrate today. That Walter Kohn, who is a physicist, who had these very special ideas on how we could understand these large groups of atoms… and it reaches all the way from solid state physics, from device physics, all the way now into biology.[379]

In his speech to the press, Walter described the difficult time of his youth in the Nazi era and the fact that "he never graduated from high school and never really from college either". He gave particular thanks to his teachers including Nohel and Sabbath in Vienna. He emphasised the crucial role of the Hauffs in England and Mendels in Toronto for taking him into their family homes. He mentioned his fine teachers at the University of Toronto including Infeld and the extraordinary experience of being supervised by Schwinger at Harvard for his PhD. He gave credit to Hohenberg and Sham for their crucial collaborations with him, and also to Parr and Becke for taking DFT to chemistry. He emphasised that his broad background in physics helped him when he became the first Director of the ITP in Santa Barbara.[379]

Walter was asked by a reporter if he was surprised to win the Nobel Prize in Chemistry and he said he was, although he was well aware that his theory was being applied extensively in chemistry. He attempted to explain DFT to the press present but they were clearly puzzled and one asked: "What use is this in medicine?" However, even to that question, Walter had a good answer as DFT was starting to be used in drug discovery.

Walter's associated institutions all wrote congratulations. The President of the University of California, Richard Atkinson, put out a press release stating: "Kohn's honour is the 12th time that a Nobel in Chemistry has been awarded to a University of California researcher." Herbert York, the Chancellor of UCSD in Walter's time, wrote: "It's wonderful to know that the people who deserve it really get it." The current Chancellor of UCSD Robert Dynes reminded Walter that he was on the San Diego faculty when he did the work that led to the Nobel Prize. There was a congratulatory letter from Barbara Lee, the California Member of the House of Representatives.

After returning home to Illinois, John Pople had sent off a brief message to Walter: "It's a huge pleasure to share this award with you! I'd like to call you directly, so perhaps you can give me a call."[379] This message can be compared to the analogous one Dirac, who studied in the same city (Bristol) and university (Cambridge) as Pople, sent to his co-Laureate Schrödinger in 1933: "I am very glad indeed to be in the company of you and Heisenberg, and hope we shall have a good time together in Stockholm."[72]

Like Walter, John Pople had been receiving a large number of congratulations. In his case the feeling was that he represented the whole field of computational quantum chemistry which had finally been recognised. Many researchers in that area felt that there was something in the award to their field that indirectly acknowledged their own contributions. Pople would repeat this sentiment in numerous statements including at the Nobel Banquet in Stockholm. Roald Hoffmann, one of the few previous Nobel Laureates in theoretical chemistry, wrote in a postcard to Pople: "I'm very happy, the community also, of that I am certain … The Swedes know how to do this, so you'll have fun."[383]

Pople was pleased to receive the congratulations from the British Ambassador Sir Christopher Meyer in Washington DC: "I know that your pioneering contribution to computational chemistry has long been recognised, and I am delighted that your distinguished career has now been capped by the most prestigious award."[383] There was a similar letter from the British Consul in Chicago, where Pople was now living. There was a follow-up to these messages. Pople had retained British citizenship and the British Foreign Office nominated him for a knighthood. This was awarded in 2003 when he became Sir John Pople.

Pople was delighted to receive congratulatory letters from his former institutions in England including Trinity College, Cambridge where he had been a student in the 1940s and a fellow in the 1950s.[383] The Economics Nobel in 1998 had also gone to a Trinity man, Amartya Sen, who had just been appointed Master of the College. A special dinner was arranged by Trinity in the next year to congratulate its two new Laureates. This was not a rare occasion as, up to and including 1998, a remarkable 30 fellows or former students of Trinity College had won a Nobel Prize. Trinity then made Pople an Honorary Fellow in 1999, and Cambridge University gave him an Honorary Degree in 2001. Furthermore, the Royal Society awarded Pople the Copley Medal in 2002, its leading award which had been won previously by Faraday, Darwin, Einstein, Planck and Bohr. Pople had, therefore, won essentially all the major prizes and was in very good company.

The British National Physical Laboratory, where Pople had briefly been Director of the Basic Science Division in the early 1960s, also wrote with some reflected glory. The Deputy Director Andrew Wallard said: "I was delighted to see that you had been awarded the Nobel Prize for Chemistry. This will give considerable pleasure to NPL staff who were either in Basic Physics or who knew you through an association with it… NPL has been through a variety of changes but is now fighting back strongly."[383]

Carnegie Mellon University had much to publicise as both Laureates had been professors there and this was the place where Pople had done his prize-winning work. Walter and Pople were both invited to come and give lectures at a celebratory event. The day after the Prize was announced, President Jared Cohen wrote to Pople triumphantly:

> Everyone at Carnegie Mellon — past and present — was thrilled to hear the news of your Nobel Prize. This is, of course, a magnificent achievement, and we salute you! Would you care to return to the scene to rehearse (or reprise) your Nobel lecture? It would be a deep honor to have you back here for a day… It is our intention to extend the same invitation to Walter Kohn, and hope it will be possible to have both of you here together.[383]

UCSB gave Walter $33,000 to cover a new secretary to deal with all the Nobel correspondence. Then, on 12 November 1998, Walter received a letter from the White House in Washington DC from President Bill Clinton:

> I am delighted to congratulate you on receiving the 1998 Nobel Prize in Chemistry. This high honor is a testament to your outstanding work in developing a theoretical framework to calculate the structures of very large molecules and to map chemical reactions. Your use of the principles of physics to advance the science of chemistry demonstrates the ever-increasing interdependence of scientific fields and the growing importance of information technology to science. You should be proud of this great achievement. I commend you for your commitment to scientific progress, and I send my best wishes for continuing success.[379]

Walter was invited to attend a dinner in his honour and address the Regents, Chancellor and President of the University of California on 19 November 1998. He emphasised how important the University of California was for him, first in San Diego (where he said he was the "second faculty and second department member") and then in Santa Barbara. As a new Nobel Laureate he also took this opportunity to criticise the Regents for continuing to be involved in running the nuclear weapons operations in the Lawrence Livermore and Los Alamos laboratories.[379] In his new more visible role, this is a topic he would come back to on a regular basis.

After the euphoria and all the congratulations, Walter had to start to make the many arrangements for the trip to Stockholm. However, a problem arose in that his wife Mara was due to have surgery which would coincide with the date in December for the presentation of the Prize. Walter then had to make the unusual request to the Royal Swedish Academy of Sciences to postpone the presentation of his Prize until the next year. This was duly granted.

The Prize to Walter was 3,500,000 Swedish Kronor which was close to $430,000. This was taxed in the USA at a total rate close to 50%. It is ironic that there is no such tax in the UK under the Inland Revenue rules for such a prize, so John Pople may have been better off if he had stayed in England after all. Walter took the advice he had received from Douglas Osheroff and used some of the prize money to cover the expenses of his guests in attending the ceremony in Stockholm. He also gave some of the award to set up a trust fund at UCSB.

Things then calmed down somewhat for Walter until 28 January 1999 when he went to Stockholm to give his Nobel Lecture at the Royal Swedish Academy of Sciences and to attend a dinner there in the evening. His lecture was entitled "Electronic structure of matter — wavefunctions and density functionals". It started with some humour:

The citation for my share of the 1998 Nobel Prize in Chemistry refers to the "development of the density functional theory." The initial work on DFT was reported in two publications: the first with Pierre Hohenberg in 1964 and the next with Lu J. Sham in 1965. This was almost 40 years after E. Schrödinger published his first epoch-making paper marking the beginning of wave mechanics... There is an oral tradition that, shortly after Schrödinger's equation for the electronic wave-function had been put forward and spectacularly validated for simple small systems like He and H_2, P.A.M. Dirac declared that chemistry had come to an end — its content was entirely contained in that powerful equation. Too bad, he is said to have added, that in almost all cases, this equation was far too complex to allow solution.[384]

He then described DFT in some mathematical detail and the progress which had been made in its implementation for both solids and molecules. He finished with the statement: "DFT has now been widely accepted by both physicists and chemists."

Walter had to wait until 10 December 1999 for the presentation of his Prize by Carl-Gustaf, the King of Sweden. His chosen guests for the celebrations were his wife Mara, his children Marilyn, Ingrid and Rosalind, his sister Minna, Pierre Hohenberg and his wife Barbara, and Robert Parr. Everyone stayed at the Grand Hotel in Stockholm. After a rehearsal in the morning, the presentations of the prizes in Physics, Chemistry, Physiology or Medicine, Literature and Peace were made in the Stockholm Grand Concert Hall. The dress code was "For gentleman, white tie and tails — a long evening gown for the ladies."[379] The ceremony started, and was to finish, with the Swedish National Anthem and musical serenades. The Chemistry Prize for 1999 was first presented to the sole winner Ahmed Zewail from Caltech, quite near to UCSB, for his work on femtochemistry. Then Walter received his prize for 1998 from the King which included the Nobel Medal and a unique certificate by the artist Bengt Landin and the calligrapher Annika Rücker (see Figure 68).

There was a brief introduction by the physical chemist Bengt Nordén who said the prize was "For your contributions to quantum chemistry, which are important also for this year's Nobel Prize as they are crucial for interpretations of experiments." In the previous year's ceremony, Björn Roos, in introducing John Pople, had given a more lengthy statement about the contributions of the two Laureates (see Figure 69). This included the words:

Fig. 68.   Walter Kohn receives his Nobel Prize from the King of Sweden on 10 December 1999. (Photograph by Hans Mehlin, 1999. Courtesy of the Nobel Foundation.)

Fig. 69.   Professor Björn Roos giving the Presentation Speech for the 1998 Nobel Prize in Chemistry. (Photograph by Hans Mehlin, 1998. Courtesy the Nobel Foundation.)

Today, we celebrate the fact that mathematics has invaded chemistry, that by means of theoretical calculations we can predict a large variety of chemical phenomena. Professors Walter Kohn and John Pople have individually made fundamental contributions to this development... Walter Kohn showed an alternative way in which quantum mechanical equations can be approximated. He showed that there is a one-to-one correspondence between the energy of a quantum mechanical system and its electron density, which is a function of three positional coordinates only and is, therefore, much easier to handle than the complicated wavefunction, which depends on the positions of all electrons. He also developed a method which made it possible to construct a set of equations, which could be used to determine the energy and electron density. This approach, called density functional theory, has developed during the last ten years into a versatile computational tool with many applications in chemistry.[385]

Walter's Nobel Medal was of 18-carat gold, plated in 24-carat gold (see Figure 70). On one side was the face of Alfred Nobel, on the other side was engraved W. Kohn MCMXCVIII together with a relief of the Goddess Isis, who has a veil held by a woman representing scientific genius. There is also

Fig. 70.   Walter Kohn's Nobel Prize medal. (Courtesy of Nate D. Sanders Auctions.)

the Latin inscription from the Aeneid "Inventas vitam huvat excoluisse per artes" which translates as "It is beneficial to improve life through discovered arts." In addition, there is the inscription "Reg Acad Scient Suec" referring to the Royal Swedish Academy of Sciences.

The Nobel Gold Medals are very valuable and several have been auctioned recently. On 10 February 2022, six years after his death, Walter Kohn's Nobel Medal was sold at auction for the substantial sum of $458,000.[386]

The great day of the Nobel Prize presentation finished with a dinner in the evening in the Grand Hotel hosted by the Royal Swedish Academy of Sciences. All the great and good of Stockholm were there to congratulate Walter and the Laureates for 1999. On the next evening the King and Queen held a magnificent banquet at the Royal Palace for the Laureates and their spouses.

Chapter Thirteen

Politics*

In an interview in 2004, when asked how the award of the Nobel Prize had subsequently influenced his life, Walter responded: "People take you more seriously ... One now has more responsibility to do something."[387] Walter already had a long history of activism in academic politics. He was someone who had seen the rise of the Nazis, suffered the Kristallnacht, taken the Kindertransport, been interned as an "enemy alien" and transported across the Atlantic Ocean. In addition, his parents, cousins and teachers were murdered by the Nazis. Walter Kohn had better reasons than most scientists to express his deeply held convictions.

For his whole academic career Walter had concerns about nuclear weapons. He had interacted with some of the famous physicists who had worked at Los Alamos and other laboratories that produced the first atom bomb. Like many of these scientists he had doubts about the future use of nuclear weapons. Back in October 1956 Walter had organised a letter to the *New York Times* from 11 physics academics at Carnegie Tech on the banning of H-bomb tests:

> In view of the widespread public discussion of proposals to ban the further testing of H-bombs, we wish to express our opinion as citizens and scientists. We are convinced that the United States should seek an international agreement to end such tests. These are our main reasons:

---

*The views and opinions of the persons recounted in this chapter are in their individual capacity and do not necessarily reflect the views or opinions of the Author of this book, or the Publisher and its employees. Your use of this chapter implies your acceptance of this disclaimer.

1) An agreement to stop H-bomb tests would be a break in the years of stalled negotiations on disarmament. In contrast to more comprehensive proposals, such a limited agreement could be put into effect immediately and would be a first step towards general arms control.

2) We believe that the security of the United States will not be endangered by our adherence to a worldwide agreement to stop these test explosions. Such a test ban would apply equally to all countries. Any violation of the agreement could be detected by monitoring devices now in use.

3) The radioactivity spread by continued tests may prove harmful to the health of people everywhere. While some scientists believe that little harm will be done, the truth is that no one really knows how great the harm may be. If tests are continued, there is a possibility that serious radioactive contamination may be produced before the true extent of the danger is known. Moreover, as the basic scientific principles on which thermonuclear bombs are based are now widely known, the number of different countries which conduct tests is likely to increase, unless all test explosions are stopped.[388]

This was the start of Walter's academic activism that would continue for the remainder of his career and would accelerate on a wide variety of issues after the award of his Nobel Prize. The management of nuclear weapons laboratories by the University of California was a particular concern for Walter. In August 1990 he had published a letter with the Berkeley professors Robert Bellah and Nobel Laureate Owen Chamberlain in the *Los Angeles Times* with the title "Weapons Labs: Light-Years From the Teaching Mission":

The University of California management of Livermore and Los Alamos is an anachronism harmful to university values... The laboratories each have about 8,000 employees and a budget of $1 billion a year, of which 75% is used for military purposes and 25% for non-military work, such as energy research. The university's net management fee is about $7 million a year. Since the end of World War II contract renewal has taken place every five years rather routinely. This time the situation is different. Last October, a special faculty committee, following an intensive 2 and 1/2 year study, concluded that UC's management of the weapons laboratories was inappropriate in peacetime and recommended that it be phased out... The state and the nation will be much better served if this great public university phases out an anachronistic management function and focuses its efforts on its primary missions of teaching and research, as well as on public service activities appropriate to a university.[389]

The letter was directed at the Regents of the University of California who were due to meet to discuss this issue. However, the Regents did not follow the recommendations to stop the management of the weapons labs at that time. As a Professor in the University of California, this was an issue that Walter would turn to several times, including when he had the golden chance to address the Regents at the dinner in his honour in 1998 after his award of his Nobel Prize. Perhaps his arguments and influence eventually had some effect as, in the period 2006–07, the management of the weapons laboratories was handed over to private companies, although the University of California still played a minor role.

In 2001, a statement was published and signed by 100 Nobel Laureates including Walter. It was titled "The Dispossessed" and had the first paragraph: "The most profound danger to World Peace in the coming years will stem not from the irrational acts of states or individuals but from the legitimate demands of the world's dispossessed."[390] However, not all Nobel Laureates who were asked agreed to sign the statement. When asked by the Chemistry Nobel Laureate from Toronto John Polanyi if he would also sign, Walter's co-Nobel Laureate John Pople had a more conservative view (on 16 August 2001):

> I agree that a statement drawing attention to the major problems facing the world in the next century is a good idea. However, I am afraid that I cannot add my signature to the suggested draft as it directly criticizes some policies of the US government that I support. Global warming is indeed significant. However, responsibility for this and other ecological concerns does not lie entirely with the rich countries. Factors such as destruction of rain forests and failure to limit the population explosion come to mind; all nations need to address them. I do favor some international control of carbon dioxide emission but not necessarily that proposed at Kyoto.[383]

He then gave his views on poverty and nuclear disarmament:

> I also feel that the proposed statement lays too much emphasis on the relation between the danger of war and third world poverty. The disaster in Nazi Germany had little to do with poverty. India and Pakistan are poor nations with nuclear weapons but I doubt that their attitude to each other would change if they became rich. I support verifiable disarmament although a good case can be made for missile defences, ultimately under international control.[383]

Pople went on to discuss the term "dispossessed":

> I find the phrase "legitimate demands of the dispossessed"
> undesirable. Attempts to improve conditions in poor countries
> can and should be based on compassion and self-interest
> rather than legitimacy. In any case the term "dispossessed"
> implies some sort of theft by the affluent. In fact the third-
> world has derived many benefits from its contact with western
> science. We should be proud of this and earnest to do more.[383]

In February 2007, with increasing tensions between Iran and the USA,
Walter, together with eleven other US Nobel Laureates and several other
prominent physicists, published a letter in the *New York Times* with the
heading "The US Congress Should Act Against Nukes":

> As physicists, members of the profession that brought
> nuclear weapons into existence, we urge the U.S. Congress
> to pass binding legislation to restrict the authority of the
> U.S. president to order nuclear strikes against non-nuclear-
> weapon states. There are no sharp lines between small
> "tactical" nuclear weapons and large ones, nor between nuclear
> weapons targeting facilities and those targeting armies or
> cities. Crossing the nuclear threshold, even with a low-yield
> weapon, would erase the 60-year-old taboo against the use of
> nuclear weapons and make their use by others more likely. If
> the victim is a non-nuclear-weapon state, such action would
> destroy, or at the very least severely undermine, the Nuclear
> Nonproliferation Treaty, with disastrous consequences for
> America and world security.

A decision that would have a major impact on the course of history and could ultimately threaten the survival of civilization should not be in the sole hands of the president unless absolutely unavoidable. We urge Congress to pass binding legislation to forbid the use of nuclear weapons by the United States against countries that do not possess nuclear weapons, except with explicit prior congressional authorization for such action.[391]

Walter was also active in discussions of the National Academy of Sciences on "Human Rights and Human Survival". In 1988 he emphasised the importance of open interchange in science and the influence of Niels Bohr on his views:

Where the main threat to peace is the confrontation between the two superpowers, I think increased mutual knowledge and an increased openness are, in the long run, at least a necessary condition for eventual disarmament and a long-term solution. I remember a position Niels Bohr took when I was a postdoc in Copenhagen. He argued that the scientists, because they are an international community, a community that naturally, because of their common interests, transcends national boundaries, have an obligation and an opportunity to be in the forefront of establishing that openness which he felt was needed much more broadly, but where scientists had special qualifications.[392]

In 1986, Walter, together with Roger Revelle, who was now back at UCSD, and the Berkeley Professor of Law, Frank Newman, organised a conference in Rancho Santa Fe, California, titled "Perspectives on the Crisis of UNESCO". This was under the umbrella of the Institute on Global Conflict and Cooperation of the University of California. UNESCO is the United

Nations Educational, Scientific and Cultural Organisation from which the US government had proposed to withdraw funding. Walter and his two colleagues stated:

> UNESCO is one of the United Nations bodies that notably promotes and supports basic research. In our view, this emphasis is essential because of the dramatic, indeed radical, changes in the world that must result from the seemingly inevitable future doubling of Earth's human population. However, long term interests in science and education will be served both by advancement of knowledge and by the provision of technical and other assistance by developed countries to less developed ones. Global programs in these areas could contribute significantly to the actual security of the United States; those objectives depend in part on intergovernmental cooperation.[393]

Walter and Revelle also wrote a letter to *Science* in 1987 proposing that Abdus Salam, who had shared the 1979 Nobel Prize for Physics with Glashow and Weinberg, become the Director-General of UNESCO.[394] However, the Spaniard Federico Mayor Zaragoza, a previous Deputy Director-General, was appointed. Walter continued with this interest and was interviewed by UNESCO TV in 2011 under the heading *Science and Peace*. He spoke in his usual style of carefully worded phrases:

> Scientists have a very good record in working peacefully with each other and overcoming national boundaries... UNESCO was a worldwide response to the horrors of the Second World War. We all owe its founders a great debt for their wisdom in choosing science as a means of strengthening international peace... I have a more important distinction than the Nobel Prize — the UNESCO Niels Bohr Gold Medal — and I am very proud of that![395]

Walter's visit to Moscow in 1963 had opened his eyes to the problems faced by scientists in Russia. He went again in October 1974 and was closely involved in the cases made for visits of Russian scientists to the USA in the 1970s. Isaak Khalatnikov was the founding director of the Landau Institute for Theoretical Physics of the USSR Academy of Sciences. He was a Jewish theoretical physicist of broad interests who had worked with Landau on the science behind the Soviet nuclear bomb. He strongly promoted interchange between Russian and American physicists. Walter had made arrangements for Khalatnikov to visit him at UCSD in 1974. Walter solicited the support of William McElroy, the Chancellor of UCSD, who wrote to the Secretary of State Henry Kissinger and Secretary of Defense James Schlesinger requesting permission for the visit. However, much to Walter's annoyance, this was denied due to the "military sensitivity of the San Diego area".[396]

Walter was always concerned about Jewish issues and had firm views on the State of Israel. In a letter to the *Prensa Popular*, a student newspaper in La Jolla, he wrote in 1974:

Once again, on browsing through the Prensa Popular, I was depressed to find a slogan-ridden anti-Israel editorial, this time entitled "Partial Israeli Withdrawal No Solution." To one who witnessed what was called "The Final Solution of the Jewish Problem" in the thirties and forties, the last words especially had an ominous ring. This apprehension was confirmed by reading the editorial itself. For, after claiming that Israel has no right to exist as a Jewish homeland, no right to try and negotiate territorial compromises to help defend itself, that it was racist and oppressed its Arab population, it ends by concluding that such a "nation… cannot survive"… Let me try and contribute a few facts and observations which I hope will somewhat correct the highly distorted picture your articles have provided.

You speak of the "myths of a historic national mission" of the State of Israel. Why myths? Israel has been the only national homeland of the Jewish people continuously for over 3000 years. During this entire period, in spite of many violent upheavals, Israel has been the home of a sizeable Jewish population and the seat of major religious and philosophical schools. Beginning in the late 1800's, Jews, whose ancestors had been driven into dispersal, began to return in substantial numbers to Israel... A large fraction of these immigrants were ardent socialists who formed communes (the so-called kibbutzim) which have been recognised as one of the most important and successful social experiments of our times.

He then went on to support his arguments through his own experiences in a visit to Israel:

So what is mythical about Israel? If you are looking for a myth I have a good one for you: The myth that all the Arabs want is to regain their ancestral lands wrongfully wrenched from them by Israel in 1967. I visited Israel in 1966... A democratic state, with reasonably equal rights for all citizens, Jews, Moslems and Christians, exists in Israel to-day... What is remarkable though, and very hopeful, is not the isolated acts of violence and repression but the unbelievable degree to which Arabs and Jews live peacefully together when given half a chance. Your editor should visit the Old City of Jerusalem, a square mile holy to three great religions, where since the city's unification four or five different worlds have become peacefully intertwined. And he should watch the Arab visitors cross the Jordan, 50,000–100,000 per year, coming from the entire Arab world to visit friends, relatives and holy places and to take a look around in Israel.[397]

Walter was an active member of the Jewish communities in Pittsburgh, La Jolla and Santa Barbara. At UCSD, he played a major role in helping to found

Fig. 71. Viktor Ullmann, in his Czech identity card. (Courtesy of Národní archiv (National Archives), Policejní ředitelství Praha II — všeobecná spisovna (Police Head-quarters Prague II. — General Register), the period 1941–1950, record signature U 109/7, box 11 856, Viktor Ullmann, born 1. 1. 1898.)

the Department of Judaic Studies. As a strong supporter of science in Israel, he was a Visiting Professor at the Hebrew University, the Weizmann Institute of Science and Tel Aviv University. Especially after his marriage to Mara in 1978, his Jewish heritage became even more important to him and he would reflect in many interviews on the fate of his parents Salomon and Gittel Kohn, and his cousins Georg and Lilly Kohn in the Holocaust. In 2015, just one year before he died, Walter submitted handwritten documents to the United States Holocaust Memorial Museum to make sure they will be remembered.

Viktor Ullmann was a distinguished composer who was born in Silesia and studied in Austria (see Figure 71). He was a student of Arnold Schoenberg. In 1942 Ullmann was sent to Theresienstadt, like Walter's parents and many other Jewish people. While a prisoner in the camp under very difficult conditions he organised concerts and composed several original musical pieces including instrumental, vocal and stage works.[398] They included a Piano Sonata and a Quartet that became highly rated. He was transported to Auschwitz on 16 October 1944 on the same train as Walter's teacher in Vienna, Emil Nohel, and just a few days before Walter's parents

Salomon and Gittel were also sent there. Walter organised and sponsored a performance of Ullmann's String Quartet No.3 in the Lotte Lehmann Concert Hall in Santa Barbara in 2010. He explained his motivation: "My parents, Salomon and Gittel Kohn, were in the Terezin [Theresienstadt] concentration camp with Viktor Ullmann. I would like to think that they heard some of his beautiful music."[399]

During 1994, Walter went to Kraków in Poland for a conference titled "Thirty Years of DFT". He was also able to view a photographic exhibition on "Kraków Jews 1868–1939" which featured the work of Roman Vishniac, the father of Walter's wife Mara. With three friends, he was taken to visit nearby Auschwitz, where his parents were murdered. He was given a detailed tour of both Auschwitz and Birkenau where the museum displays its unique collection of tragic artifacts including prisoners' hair, gold teeth, luggage, clothes and children's toys. Walter took note of all the details. On the way back to Kraków nobody spoke.[400]

Walter had strong views on his home country of Austria. After the award of the Nobel Prize in 1998 he was immediately contacted by media outlets from Austria wanting interviews. He was reluctant to accept these invitations. He often said that many Austrian people were not supporters of the Nazis before 1938 but that all changed after the Anschluss. He felt that Austrians in Vienna were very much involved with forcing him and his sister to leave the country, and in impoverishing his parents and sending them to the Theresienstadt camp at Terezin. He said he was happy for the USA, Canada and even the UK to be associated with his Nobel Prize but he did not want to be claimed by Austria.[6] Walter was particularly annoyed by the headline of the conservative newspaper in Vienna *Die Presse* which, after the announcement of his Nobel Prize, had the jubilant headline "25 years after Konrad Lorenz, Austria can once again claim a Nobel Prize Winner".[401] Lorenz was awarded the 1973 Nobel Prize for Physiology or Medicine. However, he had joined the Austrian Nazi Party in 1938, was a doctor in the German forces on the Eastern Front and conducted racial studies in occupied Poland. Thus the comparison of Walter with Lorenz was particularly insensitive.

Walter felt strongly that Austria had not properly acknowledged its role in the Holocaust. In a subsequent interview he said:

Vienna had not had a Nobel Prize for a long time. There was a newspaper headline "Austrian Chemist Wins Nobel Prize." Well, "Chemist" is understandable but "Austrian" was not right. Two weeks later I received a telegram from the President of Austria. He said "Like all Austrians, I am very happy that you received the Nobel Prize. We have heard in spite of the difficulties you have had, you have always remained committed to your fatherland." This message was infuriating given that the Austrians had arrested my parents and sent them to Theresienstadt.

I am particularly critical of Kurt Waldheim who had been President of both the United Nations and of Austria. He had been a German Army officer involved with the arrest and execution of freedom fighters in Yugoslavia, which he omitted in his biography. Waldheim said "If I am a criminal … then we are all criminals." Austria had an agreement with the Allies that they were not responsible for the Holocaust.

I sent back a critical response to the telegram from the President with my views in which I tried to be polite but clear. I immediately got back a second message that was very apologetic. I could not accept the statement (on the Holocaust): "Yes it is happening but I could not do anything." The most important lesson of the Holocaust is that you can do something, even if it is small.[6]

However, Walter was always flexible and was prepared to change his mind. After Christmas 1998, on his first trip overseas after the Nobel announcement, Walter visited Vienna at the invitation of the President of Austria Thomas Klestil and the President of the Parliament. He met with members of his family including his sister Minna, nephews, nieces, and grand-nephews and grand-nieces. He gave a lecture at the Technische Universität, where he had previously been awarded an Honorary Degree, and was taken by friends to a party at his favourite coffeehouse, the Café Sperl.[402] He was presented with the Decoration for Science and Art from the Republic of Austria for which "The number of living recipients is limited to a maximum of 72 at any one time (36 recipients for science and 36 for arts). In each of these two groups there are 18 Austrian citizens and 18 foreign nationals."[403]

During this period Walter was interviewed by *Austrian Kultur* and he spoke frankly:

> In terms of my identity, I see myself as an American, a world citizen, a Jew, and a former Austrian… I lived in Austria until the age of sixteen and I have some wonderful memories and many things that I am grateful for. For example, I feel that I got an excellent education at the Akademisches Gymnasium in Vienna. The fact of the matter is, however, that the Austrian authorities expelled me from that school in a devastating way. I then had an opportunity to continue in a Jewish school… I managed to get out of Austria on a Kindertransport to England three weeks before the war broke out. I left without my parents, who I know went via Theresienstadt to their death in Auschwitz. There were people in England, in Canada, in the US who instead of trying to eliminate me, really supported me. With all that in my mind, when people say "Hooray for an Austrian Nobel Laureate" I have problems.[404]

On the day after the Nobel announcement, Erich Lorenz, Chair of the Parent's Association of the Akademisches Gymnasium in Wien, had sent an email message to Walter:

> Congratulations on being awarded the Nobel Prize! It's a shame that I only now found out that you had to leave the academic high school at the time. This year we — students, parents and teachers — held a memorial service at school with former students who suffered the same blatant injustice, as a reflection

and warning for today's youth. I am pleased, as I studied at the TU Vienna, that your thoughts are valued there and in Vienna. May I ask you — when you are back in Vienna — to address some words to the students in the Akademisches Gymnasium? You as a role model could tell them what is important in life![379(t)]

On his return to Vienna in January 1999, some 60 years after his expulsion, Walter recalled this message and went back to the Akademisches Gymnasium, but with some trepidation. A description on this visit stated:

It was quite obvious that the school wanted to show off with its famous son, with a Nobel laureate. Walter admitted that — at first — he was a bit upset but then he accepted and went to the school, where they showed him around. Among other things he attended a history class, in which the teacher had started a project on the Nazi-period looking at specific family tragedies. The pupils started conversations with Walter and asked him many questions about that time. It was these discussions that made him see Austria differently, in a way that was new to him. Now comes the unexpected reaction of Walter. He changed his mind and made a significant donation to the school in order to continue such efforts. It takes a great personality to donate so generously to the school that once expelled him.[405]

Walter established a foundation which each year would honour a student's work at the Chajes and Akademisches Gymnasiums in the natural sciences and in the humanities.[405,406] The prize was set up in honour of his teachers at the Chajes Gymnasium, Emil Nohel and Viktor Sabbath. It would be awarded by an external jury which included academic colleagues from Vienna. Even before the announcement of his Nobel Prize, a plaque had been placed in the Akademisches Gymnasium with a list of the Jewish students who were expelled in 1938.

Eric Kandel is another Austrian scientist who was forced to leave Vienna after the Anschluss. He went to the USA in May 1939 at the age of nine, had a very distinguished career in biomedical sciences at the New York University Medical School and Columbia University, and was awarded the Nobel Prize for Physiology or Medicine in 2000 for his research on signals in the brain.[407] Kandel had similar views to Walter on Austria. Like Walter, he received the congratulations from the President of Austria Thomas Klestil. Kandel suggested to the President that they should organise a symposium in Vienna for "comparing Austria's response to the Hitler period, which was one of denial of any wrongdoing, with Germany's response, which was to try to deal honestly with the past." This symposium was organised at the University of Vienna in 2003 and both Walter and Kandel spoke candidly about their experiences. In his biography, Kandel gave a related anecdote that illustrates Walter's activism:

> While we were in Vienna in June 2003, Walter Kohn and I learned that the Viennese Kultusgemeinde, the Jewish social service agency that is responsible for the synagogues, the Jewish schools and hospitals, and the Jewish cemetery in Vienna, was going bankrupt trying to protect those entities against continuing vandalism. European governments typically compensate Jewish agencies for such expenses, but the Austrian government's compensation was not adequate... Back in the United States, Walter Kohn and I joined forces to see whether we could help ameliorate the situation. Walter had gotten to know Peter Launsky-Tieffenthal, the consul general of Austria to Los Angeles, and Launsky-Tieffenthal arranged for a conference call that would include himself, Muzicant [the agency president], Wolfgang Schüssel (the chancellor of Austria), Walter, and me. We thought the conference call was all set, but at the last moment Schüssel cancelled. He did so for two reasons. First, he was concerned that his participation might be taken as an indication that the Austrian government was not doing enough for the Jewish community, which he denied. Second, he was willing to speak to Walter Kohn but not to me, because I had been critical of Austria.

Fortunately, when Walter and I were in Vienna for the symposium, we had also met Michael Häupl, the mayor of the city of Vienna and governor of the state of Vienna. We were very impressed with Häupl, a former biologist, and greatly enjoyed our evening with him. He acknowledged that the Jewish agency was being short changed. After Schüssel refused to talk to us, Walter wrote Häupl, who swung into action below the federal level. To Walter's and my delight, he succeeded in persuading the governors of the Austrian states to help out financially.[407]

In due course, Walter received numerous further honours from Austria. In 2008 he received the Great Silver Medal of Honour with Star for Services to the Republic of Austria.[403] Then, in 2012, he received an Honorary Doctorate in Science from the University of Vienna alongside the chemical entrepreneur, Alfred Bader, a Kindertransport refugee sent to Canada on the *Sobieski* with Walter, and Peter Pulzer.[408] Pulzer had emigrated from Vienna to England in 1939 at the age of nine. He was educated at Cambridge University and eventually became Gladstone Professor of Government at All Souls College, Oxford. This award to Kohn, Bader and Pulzer was intended to indicate the sincere regret of the University of Vienna for what had happened in the late 1930s and its hope for the future. The University of Vienna had seen dark days after the Anschluss when as many as 322 professors were expelled from a total of 763. One of those expelled was Erwin Schrödinger.[72]

Walter continued his interest on matters related to Vienna into his 90s. Together with the Holocaust survivor Elie Wiesel, who won the Nobel Prize for Peace in 1986, Walter was made an Honorary Member of the Board of the *Vienna Project*. This project established the first public art memorial in Vienna to represent the many groups of Austrian victims murdered by the Nazis. The project "forged a dynamic relationship between performance art, video and new technologies, typography and street art, web design, history, archival research, and Holocaust education."[409]

Walter not only had strong views on the role of Austria in the war but also the Catholic Church. In February 2000, he received an email message from Giovanni Bachelet, a professor of physics of the Sapienza Università di Roma who had taken part in an ITP workshop:

I am writing you to inquire about your participation to the International Conference "Physics for the 21st century", to be held in Rome, Italy, on September 6, 7 and 8. This conference is part of the celebrations for the Great Jubilee of the Catholic Church and I was one of the consultants who contributed to the definition of its program. The Organizers have told me that the official invitation should have reached you by now, and asked me if I could inquire about the possibility that you accept such an invitation and give your invited talk in Rome next September. So here I am. I am well aware that you are full of work and invitations, and actually I have just seen that you might already be in Italy in the same days because of another conference in Sardinia; I do not know whether this coincidence will increase or decrease the chance that you speak in Rome too. Also, I do not know what your feelings may be about churches, popes, and the like. My personal contribution to this program, as a church member, was the attempt to have top speakers who can give a perspective of science rather than promoting a conference of "Catholic scientists".[410]

Walter replied at once with some criticism of the Vatican:

Thank you for the kind invitation which I had not yet seen. I was very disappointed that the Vatican did not use the occasion of the Great Jubilee to face its great failure during the Holocaust, some of which I experienced first-hand (e.g. the Viennese Cardinal Innitzer receiving Hitler's envoy with the Heil Hitler salute). My parents and others close to me were murdered. As you know, I am Jewish, but raised in a very ecumenical spirit. My closest boyhood friend and, of course, many dozens of my adult friends have been Christian. I believe you will understand that I cannot participate in a Vatican-sponsored Jubilation under these circumstances. But I have a suggestion: Please consider compiling a list of Jewish physicists who became victims of the Holocaust (including my superb high school teacher Emil Nohel, who had been an assistant of Einstein's), and bringing it to the attention of the participants.[411]

However, the persuasive Bachelet would not take no for an answer and they had further discussions. In the middle of their debate, a month later CNN reported: "Pope John Paul sought forgiveness on Sunday for the many past sins of his Church, including its treatment of people of other religions."[412] The Pope then visited the Holy Land for the special year of 2000 where he also expressed these sentiments. These timely comments softened Walter's views and he wrote to Bachelet:

After considerable soul searching, I concluded that I should accept your invitation. I will now begin trying to obtain the names and particulars of physicists (and perhaps also chemists) who lost their lives during the Holocaust and I will also make some brief remarks about the tragic, widespread silence during the Holocaust and the need to speak up vigorously and fearlessly to help avoid future catastrophes. I expect that all those remarks can be offered in no more than 10 minutes. If the list of physics and chemistry victims should be very long, I would plan to distribute it in printed form rather than in my verbal remarks.[413]

Bachelet wrote on Walter's visit to the Vatican:

> In the end Walter came and gave a wonderful talk exactly in these terms. Rather than blaming anyone for failing to speak out, he asked the audience a constructive and much more poignant question: What are you going to do next time? The conference was followed by a large papal audience in the Vatican, at the end of which Walter was one of the few who were admitted to see the Pope, while most of us watched them from a distance on a large TV screen... In his own recollection, Walter's words had been something like, "Your Holiness, I appreciated all your efforts to better understanding between the Christian and Jewish people" to which the Pope had smiled and replied with a repeated "Thank you".[414]

This was not the end of Walter's interaction with the Catholic Church. He then had a correspondence with Józef Życiński, the Archbishop of Lublin, whom he had met at the Vatican. In an earlier part of his career Życiński had taught the philosophy of physics at the Pontifical Academy of Kraków where he had studied some of Walter's papers. He also had a further connection with Walter as he had been Bishop of Tarnów, an area which included Brody in Galicia where Walter's mother Gittel was born. The letters between Życiński and Walter went quite deeply into the relations between science and the Catholic Church with Galileo inevitably being a subject of debate.[415] Bachelet was involved in these discussions and, in a letter to him, Walter mentioned a charming anecdote about a visit he had once made to Venice:

> I came downstairs to the hotel restaurant for breakfast, with my wife joining me shortly thereafter. She struck up a conversation with an Italian lady and her daughter sitting next to us. The lady remarked that she had noticed me when I began my breakfast alone, and had reached the conclusion that, if I would be joined by a woman, I would probably be a theoretical physicist, and if not, I was probably a priest.[416]

These discussions sparked an interest of Walter in the relationship between science and religion. He gave a long presentation on this topic, and his recent interaction with the Catholic Church, in a video for the UCSB programme "Science, Religion and Human Experience".[417]

Walter had an interest in solar power and climate change and gave many lectures around the world on these topics in his later years. In 2005 he was involved with an original educational project on solar energy. His film *The Power of the Sun* gives the scientific story of photovoltaics, including a description of the fundamental ideas on light from Newton and Einstein to the invention of practical solar cells at Bell Labs in the 1950s, and then the development of the latest solar panels. The narrator was the popular British actor John Cleese, who had taken up residence in Santa Barbara.[418] The film was made in collaboration with John Perlin, a colleague at UCSB.

Alan Heeger was also an advisor for *The Power of the Sun*. A professor in the physics and materials science departments at UCSB, he won the Nobel Prize for Chemistry in 2000 for discovering electrically conductive polymers.[419] So Heeger was the second physics professor at UCSB to win the Chemistry Prize — a remarkable coincidence. To date, UCSB claims six Nobels and Walter was the first. In addition to Kohn, Heeger and Gross, they are the electrical engineer Herbert Kroemer, who won the Prize for Physics in 2000 for his work on semiconductor heterostructures, Finn Kydland, who was awarded the Economics Prize in 2004 (and had also previously been on the faculty of Carnegie Mellon University), and Shuji Nakamura, a professor of engineering and materials, who won the Physics Prize in 2014 for discovering blue light-emitting diodes.[420] Apart from Gross, it is interesting that Kohn, Heeger, Kroemer and Nakamura won the Nobel Prize when working in departments not directly concerned with the title of the Prize. This emphasises the power of the interdisciplinary research at Santa Barbara.

Chapter Fourteen

Legacy

As we have seen from the previous Chapter, Walter turned his interests to numerous global issues after the award of the Nobel Prize in 1998. As a Research Professor of Physics at UCSB he did not now have formal duties for teaching, and had more time to pursue his wider interests. Nevertheless, he did publish a small number of research papers after 1998 in which he addressed effects such as degeneracy, symmetry and magnetism in DFT.[421,422,] He also wrote some reviews and obituaries, including one for his great friend and collaborator Luttinger who had died in the year before the announcement of Walter's Nobel Prize.[130]

After 1998, Walter was much in demand to speak at conferences and, inevitably, many of these were in chemistry. During his career in theoretical physics, he had keenly attended numerous meetings of the American Physical Society. Now he was quite frequently invited also to the meetings of the American Chemical Society, where he would wear his distinctive beret. I recall meeting him at the American Chemical Society Annual National Meeting in Chicago in March 2007. I was at the Meeting to make a presentation to David Buckingham of the Ahmed Zewail Prize in Molecular Sciences, which was awarded through the journal *Chemical Physics Letters* which I edited. Together with Zewail, I came across Walter at the meeting. He was there to present his film *The Power of the Sun*. He was very pleased to meet up again with Zewail, eight years after they had both been presented with their Nobel Prizes for Chemistry in Stockholm by the King of Sweden. Walter said to me, with a cheeky glint in his eye, that he would also like to be awarded the Ahmed Zewail Prize.

Starting, very appropriately, in Paris a biannual DFT conference was set up which took place in different countries around the world. Walter attended several of these meetings. His last one was in Athens in 2011 where he was

Fig. 72. Walter Kohn in Athens, 2011. (Courtesy of Karlheinz Schwarz.)

still able, at the age of 88, to climb the Acropolis (see Figure 72).[423] He also enjoyed speaking at conferences which had sessions on Science Policy issues. I recall a conference on *Frontiers in Molecular Science* held in Qingdao, China in July 2002. In addition to Walter, the Chemistry Nobel Laureates Rudolph Marcus, Richard Ernst and Robert Huber spoke at the meeting. After Walter's presentation on how to include van der Waals forces in DFT, I spoke on the related topic of Monte Carlo calculations on water clusters and biomolecules. However, it was in the session on Science and Society in which Walter really came alive. David King from Cambridge spoke on his role as the Chief Scientific Advisor to the Government when mathematical modelling was used to predict the spread of the foot-and-mouth disease between animals in the UK. Walter spoke with great enthusiasm on this work and stated that it was vital for governments to use scientific advice.

Walter also gave numerous named lectures, often associated with the leading scientists he had known and the institutions he had been involved with. In 2009 he gave the first John A. Pople Lecture at Carnegie Mellon University which was set up after the death of Pople in 2004.

Walter had a particular interest in Erwin Schrödinger. This was inevitable as Walter had developed in 1964 one of the most impactful modifications of wave mechanics since Schrödinger's paper in 1926. His form of DFT was Walter's great scientific legacy. Right at the start of his Nobel Prize Lecture, Walter wrote that he developed DFT "almost 40 years after E. Schrödinger published his first epoch-making paper marking the beginning of wave-mechanics."[384] Furthermore, like Walter, Schrödinger was from Vienna, was a Nobel Laureate and they had even gone to the same school, the Akademisches Gymnasium. Schrödinger was not Jewish but, like Walter, he had to escape from the Nazis. This was in September 1938 after he was fired from his professorships at the Universities of Graz and Vienna. Also, like Walter, Schrödinger then first sought refuge in England but did not stay very long.[72] There are, therefore, many parallels between Walter and Schrödinger, and Walter often discussed this in interviews. Walter said:

> I have enormous admiration for Schrödinger. My own work is entirely built on Schrödinger's work. It transforms it and looks at it in a different perspective... I consider him as my scientific ancestor... Schrödinger's equation describes all aspects of the physical world we experience on earth... Words like miraculous and preposterous come to mind.[24]

Schrödinger worked at the Dublin Institute for Advanced Studies from 1940 to 1956, and then at the University of Vienna until his death in 1961. Walter did meet some of the great pioneers of quantum theory including Bohr and Pauli but he stated that he never met Schrödinger.[24] Walter lectured at the Dublin Institute in March 1958 at the invitation of John Synge, whom he had known in Toronto when he was a student, but this was two years after Schrödinger's departure from Dublin for Austria. Furthermore, before Schrödinger's death in 1961, Walter returned to Austria on only a very small number of occasions.

During a trip to England in 2005, Walter visited the author of this book at Magdalen College in Oxford, where Schrödinger had been a Fellow from

1933–36. Just hours after his admittance as a Fellow on 9 November 1933, Schrödinger heard in the office of the President of the College that he had been awarded the Nobel Prize for Physics with Paul Dirac. I was President of the College from 2005–20 and so a visit from Walter to discuss Schrödinger was of interest to him and also to me as I was accumulating archive material for my biography *Schrödinger in Oxford*.[72] In addition, in the President's Lodgings at Magdalen College there are some fine medieval tapestries, similar to those Walter would often go to see in the Musée de Cluny in Paris.

I expected Walter to be interested in discussing Schrödinger's time in Oxford where he wrote the first paper on *Entanglement* and also his famous paper on *Schrödinger's Cat*.[424,425,] However, apart from the tapestries which he much enjoyed, the main issue Walter wanted to discuss was Schrödinger's escape from the Nazis in Austria in 1938. His view was that the Austrian people were guilty of supporting the Nazis after Hitler took over their country and brought forward laws against Jews and dissidents at an alarming speed. Accordingly, Schrödinger had to leave Austria within six months of the Anschluss and Walter departed in just over a year. Schrödinger and Kohn came up with the main working equations that are used in modern calculations on molecules and materials. There are so many similarities between them.

Walter was the first signature in the new Visitors Book at the Magdalen President's Lodgings (on 20 November 2005). Other subsequent visitors with signatures in the book included Queen Elizabeth II and the Duke of Edinburgh, Charles the Prince of Wales (now King Charles III), Prince William, Prince Edward, the Dalai Lama, Hamid Karzai (then President of Afghanistan), Hillary Clinton, David Attenborough, and several other Nobel Laureates including President Santos of Colombia (Peace), Seamus Heaney (Literature), Ahmed Zewail (Chemistry), John Polanyi (Chemistry) and Eric Kandel (Physiology or Medicine) who had organised the symposium in Vienna at which Walter had spoken in 2003.[426] So Walter was in good company here.

Four years before visiting Magdalen College, Walter had received an Honorary DSc from the University of Oxford. This was an honour not granted to Schrödinger, but it was to Einstein in 1931. The oration to Walter at the traditional *Encaenia* ceremony held in Christopher Wren's Sheldonian Theatre was in Latin and ended with the phrase: "Praesento hominum physicorum praeclarissimum, chemicorum ingeniosissimum, qui et ipse res plurimas illuminavit et ceterorum ingenia prudentissime direxit, Gualterum Kohn, Societatis Regiae Sodalem. Praemio Nobeliano nobilitatum, ut admittatur honoris causa ad gradum Doctoris in Scientia" which translated into English is: "I present Walter Kohn, Member of the Royal Society, Nobel Prizeman, an outstanding physicist who is also a brilliant chemist, a man who has made important discoveries himself, and who has directed the research of others with accomplished skill, for admission to the honorary degree of Doctor of Science."[427] Walter was amused by these words (especially "brilliant chemist") as can be seen from Figure 73.

Walter was to receive as many as 18 Honorary Degrees.[423] The first one from the University of Toronto in 1967 was very dear to his heart as

Fig. 73.   Walter Kohn listening to the oration for his honorary doctorate at the University of Oxford, 2001. (Photograph by Tony Hudson. Creative Commons Attribution-Share Alike 3.0 Unported licence.)

that was the university which started him off on his outstanding academic journey. He also much appreciated the Honorary Doctor of Science in 2012 from Harvard. This was the university that had awarded Walter the crucial scholarship back in 1946 that enabled him to do the research on quantum scattering theory that led to his PhD in just two years. He also was pleased to receive an Honorary Degree from Carnegie Mellon University in 1999, where he had his first academic appointment at Carnegie Tech. In 2011, he was elected an Honorary Member of the Austrian Academy of Sciences, an academy distinguished by managing to retain its independence from the German academies during the war.[428]

A particularly poignant prize that went to Walter in 2002 was the Prix des Trois Physiciens awarded by the Laboratoire de Physique at the École Normale Supérieure in Paris, where Walter discovered his form of DFT.[429] This prize is given in memory of the three directors of the Laboratory, Henri Abraham, Eugène Bloch and Georges Bruhat, who all died in Nazi concentration camps (the first two in Auschwitz). The prize had been awarded previously to Mott and Oppenheimer.

In 2003, Walter heard that his sister Minna had died on 18 February at the age of 83. Minna had divorced Franz Pixner, who in due course went on to marry, as his fifth wife, Alma Zsolnay, the granddaughter of the composer Gustav Mahler. Minna had continued to be involved with the Brüder Kohn collection of artistic postcards until she died. She had been very close to Walter, especially as a girl and a young woman, and they had stayed in the house of the Hauff family in Sussex during 1939–40. After her return to Austria in 1945, Walter had met his sister a few times on his visits to Vienna and elsewhere in Europe. She had attended the presentation of his Nobel Prize in Stockholm in 1999. On her gravestone in the Jewish cemetery in Vienna it states (translated) "A life in art and culture, and love for your children" (see Figure 74). It is also engraved on the stone that her parents Salomon and Gittel Kohn, her uncle and aunt, Alfred and Hilda Kohn, and her cousins Lilly and Georg Kohn were murdered by the National Socialists. In addition, it is stated that her uncle Adolf Kohn fell (in World War I) in 1918.

Minna's daughter Lisi Rosenhek wrote about the difficulties faced by her grandparents Salomon and Gittel Kohn in Vienna in the war and the

Fig. 74. Gravestone of Minna Pixner (née Kohn). (Creative Commons Attribution-Share Alike 4.0 International licence.)

courage they showed in Theresienstadt.[41] Her son Raphael Rosenhek is a distinguished heart surgeon and he has written how he was inspired by the scientific achievements of his great-uncle Walter Kohn to undertake research in cardiology.[94]

Walter had an important influence on the status of research in solid-state physics in the USA. This was an area that for many years was not considered seriously by some theoretical physicists, especially those who worked in elementary particle physics, who considered their area of research to be more fundamental. There were several pioneers on the quantum theory of solids who, despite numerous nominations, did not win the Nobel Prize for their contributions to that field including Sommerfeld, Peierls and Bloch (although Bloch was awarded the prize for his contributions to Nuclear Magnetic Resonance). Walter's papers on solid-state theory that were not involved with DFT, including his work with Luttinger on electrons in periodic fields, the Kohn-Rostoker paper on periodic lattices, and his own papers on analytic Bloch waves, Wannier functions and insulating states, continue to be influential and receive many citations.

The start of solid-state physics as a major subject can be identified with the pioneering research done at Bell Labs that led to the transistor and other electronic devices in the late 1940s. Due to the demand for research in this field, the American Physical Society created a Solid-State Division in 1947.[430] With research in fluids and phase transitions also becoming important the term Solid-State Physics was eventually replaced by Condensed Matter Physics in 1978 and this quickly became the largest division of the American Physical Society with over a quarter of its members. 1978 was also the year for the award of the Nobel Prize for Physics to Peter Kapitza for his work on superfluid helium, which was closely linked to the research of Don Misener who had taught Walter at the University of Toronto. Experimental condensed matter physics had a long history and had been the subject for one of the early notable Nobel Prizes — that to Heike Kamerlingh Onnes in 1913 who discovered superconductivity.

There were then Nobel Prizes in Physics awarded to Wilson in 1982 for his theory of phase transitions and de Gennes in 1991 for theories of complex matter. However, theoretical research in the solid-state or condensed matter field was often referred to in somewhat derogatory terms by physicists in other areas. For example, Murray Gell-Mann famously described it as "squalid-state physics" and the always sarcastic Pauli used the word Schmutzphysik (translated as "dirty physics").[431]

Their objection was that solid-state physics did not have fundamental principles from which results of experiments can be explained, unlike, for example, the Standard Model of particle physics. This view was held despite the fact that solid-state physics was having considerably more applications to people's lives than other areas of physics through the inventions of the transistor, integrated circuits, and many other breakthroughs in microelectronics. Furthermore, a whole new industrial arena was being created from modern products, including computers and mobile phones, which contained essential solid-state components. There were also major intellectual challenges. Anderson even wrote an article in *Science* titled "More is different" highlighting the diversity and complexity of effects in condensed matter physics and the need for new theories and mathematics to explain them.[432]

Things got rather bitter in the 1980s when the high-energy physicists put together a proposal to build a Superconducting Supercollider in Texas at a significant cost. This was objected to by programme leaders in other areas such as materials science where there was a concern that their budgets for research could be reduced significantly if the Supercollider plans went ahead. Anderson's view that the field of condensed matter was an endeavour that competed with high-energy physics for originality in its theory, while being a fraction of the cost in its experiments, was influential and he expressed this opinion to the US Congress. The Supercollider project was eventually cancelled in 1993.[433]

Walter made an important contribution by bringing together the warring factions of theoretical physics through his leadership as the first Director of the Institute for Theoretical Physics at Santa Barbara. He set up a well-balanced programme of workshops right across different areas of physics which attracted the very best theoreticians from around the world. There was an emphasis on interdisciplinary interactions between the different subfields of physics and other sciences. His warm character, his broad interests and his wide contacts were crucial to this success. This is exemplified by the comments that the future Nobel Laureate David Gross made to the press conference on the day Walter won his Nobel Prize in 1998. Just the year before, David Gross had become Director of ITP and had been a strong supporter of the Supercollider project (see Figure 75). He said:

Walter was the first and founding Director of the ITP and what we do here under this institution, which was created by Walter, is to bring together leading scientists in physics from throughout the world to tackle major problems in theoretical physics and related fields, such as chemistry, engineering, biology and the like. As the current Director of the ITP, I am continually impressed by the brilliant structure that

Walter and his colleagues put together in founding this Institution. This, what was created here, became rapidly one of the leading research institutes in physics throughout the world. It played a leading role in development of science at Santa Barbara, which has now risen to the heights of being ranked by *Science Watch* in the latest ranking as No. 1 in the nation, and has been widely copied at, I think, in something like 10 separate places throughout the world in which people have tried to imitate Walter's invention of a national institute in physics, and people have even used his model for other things such as mathematics. The founding of the ITP will undoubted be, in addition to Walter's scientific work, a lasting memorial to his creativity and genius.[434]

Fig. 75.   David Gross, 1995. (Courtesy of American Institute of Physics, Emilio Segrè Visual Archives, Physics Today Collection.)

Thus the ITP in Santa Barbara is a very significant legacy of Walter Kohn, in addition to his major scientific research contributions. It is not only Santa Barbara which should be mentioned here. Walter played a key role in the foundation and expansion of the University of California, San Diego. There are few people who have contributed more to the enhancement of the scientific reputation of the University of California than Walter Kohn.

As people get older, nostalgia sets in and Walter was always delighted to hear from his school friends who were at the Chajes Gymnasium with him in Vienna in the late 1930s. He had managed to keep in contact with some of these old friends, although, sadly, several had passed away by the time of his Nobel Prize announcement. When she heard that news, Gertrude Ehrlich said that she thought it must be another Walter Kohn as the Walter Kohn she knew had never studied chemistry.[116]

A reunion for the small number of classmates still living was organised by Walter in 2000 in Capital Hill, Washington and included Gertrude Ehrlich, Herbert Neuhaus, his wife Brigitte, and Ilse Arnold-Levai, some 61 years after they had studied together in Vienna (see Figure 12 for what they looked like at the age of 15). Walter had also hoped to meet up with the Swedish-based mathematician Karl Greger around the time of the Nobel Prize ceremonies in Sweden but, sadly, he had already passed away. However, during a visit to Trieste in 2001, Walter managed to catch up with the mathematician Rudi Permutti just a few months before he died.[25]

In recent years there has been a resurgence of interest in the remarkable Kindertransport scheme.[27,33–37] This brought to the UK close to 10,000 Jewish children from Germany, Austria, Czechoslovakia and Poland in 1938 and 1939. Films such as *One Life* featuring Nicholas Winton, who saved nearly 700 children from Czechoslovakia just before the start of the war, have been acclaimed. Over 1,000 of the children on the Kindertransport became old enough to fight in the war. Many, like Walter, lost their parents in the Holocaust. Several remained in the UK, although some returned to their home countries after the war.

Walter was not the only child on the Kindertransport to Britain to win a Nobel Prize. Arno Penzias had left Germany at the age of six to come to the UK. He worked at Bell Labs when Walter was a consultant there, although in a different section. Penzias and Robert Wilson observed radio noise in an ultrasensitive microwave receiver they had built. In due course this interference was identified with the background radiation of the universe, and Penzias and Wilson received the 1978 Nobel Prize in Physics for their unexpected discovery.[435] At the 54th meeting of Nobel Laureates in Lindau, Germany in 2004 both Walter and Penzias spoke on the same day.

In 1940, over 1,000 of the Kindertransport children were interned in Britain being, like Walter, aged 17 or over. Of these, nearly 400 were shipped to Canada. There have been many questions raised over the need to intern Jewish people who, although technically German nationals, were clearly opposed to the Nazis.[64,436] In Canada there has also, more recently, been several publications and discussions about the internment camps, including the Ripples camp in New Brunswick where Walter was sent.[51,437] Walter Kohn is often mentioned nowadays in articles on the Kindertransport and Internment.[37,51,64,437]

In his last few years, Walter became interested in an entirely new research project that had a personal aspect. His wife Mara had developed the eye disease macular degeneration. This age-related disease is quite common and leads to blurred vision due to damage to the macula at the centre of the retina. Walter wanted to develop a method to correct for this blurred vision. His idea was to use the Amsler grid, which is a graph paper pattern normally used to diagnose macular degeneration through the observed distortion of the grid of lines in the graph. Walter's clever idea was to use computer software to design an optical material to minimise this distortion. He even got technicians in the machine shop at UCSB to make such a material in 2012 to use in contact lenses or spectacles. He gave presentations in lectures and videos on this idea and it was reported in several journals on eye diseases.[438]

Walter also became known for famous quotes about science and life. "Physics isn't what I do — it is what I am" is probably the best known. Other sentiments included: "If a theory cannot be described in simple terms, it is either not a good theory or it is not well understood" and "You can never prove something correct but only test it endlessly to find a proof which shows it is wrong."[439] His special love for Paris, where he did his first work on DFT, is beautifully summarised in his quote: "Paris somehow lends itself to conceptual new ideas. I don't know why it is. There is a certain magic to that city." His comment on his departure from Vienna when he last saw his parents was truly poignant: "We would be separated, quite likely forever."

Fig. 76. Walter Kohn being interviewed in April 2014. (Courtesy of Department of Special Research Collections, UC Santa Barbara Library, University of California, Santa Barbara.)

He gave several interviews in his last few years with these messages (see Figure 76).[23]

In recent years computational developments and applications of DFT have continued unabated. The widespread use of the method in chemistry has been extended also to materials science and biology. By March 2024, the Hohenberg-Kohn and Kohn-Sham papers had received over 41,000 and 50,000 citations respectively, according to the *Web of Science*. As is often the case in science, the first papers eventually become so well-known that they are not cited by all researchers who are using the theory. Accordingly, other papers that followed on and brought DFT into chemistry and materials science have become the most highly cited papers.[440] The 1996 paper by Perdew, Burke and Ernzerhof entitled "Generalised gradient approximation made simple" now has over 152,000 citations.[441] The number for the paper by Lee, Yang and Parr of 1988 "Development of the Colle-Salvetti correlation energy formula into a functional of the electron density" is 88,000.[305] Two papers by Becke, "Density-Functional thermochemistry: role of exact exchange" (1993) and "Density Functional exchange energy approximation with correct asymptotic behavior" (1988), have over 73,000 and 46,000 citations, respectively.[306,442] There have even been scientific papers written purely about citations of DFT, such is this phenomenon which is probably unparalleled in any other area of modern science.[440]

Since the Nobel Prize in 1998 there have not yet been other truly fundamental breakthroughs in DFT but there has been a considerable amount of algorithm development, improvements of functionals and applications throughout the sciences.[443] DFT has been incorporated in numerous software packages which can be run on inexpensive computers and even laptops. Some packages are open-access but some are commercial. Indeed, the company running Gaussian eventually became so commercially successful that it split from Pople who signed a non-competitive agreement. There were also reports that some of his collaborators were "banned" from using the Gaussian software due to the possibilities of competition.[444] Former students of Pople then introduced a rival software package called Q-Chem, and early in 1999 Pople joined as a director.[445]

In 2024, DFT calculations are being done routinely by computational and experimental molecular and material scientists around the world. Many papers across the sciences frequently include pictures of molecules and chemical reaction pathways calculated using DFT. Reports of these computations supporting experimental results have become as common as X-ray or Nuclear Magnetic Resonance studies used to confirm molecular structures. Quite often, these calculations are now referred to as "first principles" with no mention of DFT or references to the original papers. The applications are not just in chemistry. There are also many numbers of papers in materials science and materials engineering where the properties and relative stability of new materials are of particular interest and this information is provided by DFT.[446] DFT calculations have led to the design of new metal catalysts for producing ammonia from nitrogen and hydrogen.[447] DFT has even been used to help predict the temperatures at the centre of the earth and in planets.[448]

In problems of biomolecular science and drug discovery DFT is now playing a major role where, for example, the free energies of small molecules binding to proteins are required.[449] As examples, DFT was used to predict new drugs used in the fight against COVID-19 and the method has also been applied in 2024 to assist in the discovery of new antibiotics.[450,451] Furthermore, the huge explosion in machine learning and artificial intelligence is helping to provide improved exchange-correlation functionals that can give reliable results for an increasing range of more complicated nano-molecular and condensed-phase systems.[452] One paper even compared the quality of 200 such functionals.[453]

Since the Prize to Kohn and Pople in 1998, there has been, to date, only one year in which Nobels were awarded purely to computational chemistry. This was in 2013 when the Nobel Prize for Chemistry was awarded to Martin Karplus, Michael Levitt and Arieh Warshel "for the development of multiscale models for complex chemical systems". Like Walter, Karplus was also born in Vienna. With his family, he left Austria straight after the Anschluss in 1938 at the age of eight.[454] At Harvard University, he carried out the first molecular dynamics simulations of the motion of atoms in proteins. Levitt and Warshel developed a Quantum Mechanical–Molecular Mechanics method that combined wave mechanics with classical molecular mechanics and allows simulations of complex problems in materials science and biochemistry to be carried out. DFT can be used in the quantum mechanical part of their theory.[455]

Since 1998 there have been just two years of Nobel Prizes for Physics awarded purely to research on the theory of condensed matter. In 2003 the Prize went to Abrikosov, Ginzburg and Leggett "for pioneering contributions to the theory of superconductors and superfluids". They were all well-known to Walter who had won the UNESCO Niels Bohr Gold Medal with Ginzburg. Then, in 2016, David Thouless, Duncan Haldane and Michael Kosterlitz became Nobel Laureates for using wave mechanics to produce new topological concepts for condensed matter physics. Their research led to the new field of topological materials that have unusual energy states and phase transitions associated with electronic structures with different characteristics to those found in metals and normal insulators. In the 2020s, DFT is being used extensively to predict new topological materials.[456]

Walter Kohn died of jaw cancer at his home in Santa Barbara on 19 April 2016 at the fine age of 93. There were then many obituaries published throughout the world which described his remarkable life and scientific breakthroughs from different viewpoints. The *New York Times* emphasised that he had won the Nobel Prize for Chemistry even though he had only studied the subject at high school. Walter's clever quote "Physics isn't what

I do — it is what I am" was also mentioned as was his efforts to terminate the management of nuclear weapons laboratories by the University of California.[457] The *Washington Post* described in some detail Walter's harsh life experiences as well as his decisive scientific contributions. It stated: "Far from home and family, Dr Kohn found himself with little to support him over a long period but his intellect and the goodwill of many strangers and mentors... At various times in his youth and early adulthood, he also worked on a farm, cut timber and prospected for gold... He told the *Los Angeles Times* that his contributions to science were his way of trying to help live his lost family's lives".[458] The French national newspaper *Le Monde* also emphasised that Walter formulated DFT in Paris.[459]

UCSB put out a press release with several quotes from his colleagues. David Gross said Walter was "a great scientist, a compassionate humanitarian and an inspired scientific statesman. As director of the KITP I learned to appreciate the genius who created the essential features of this institution that he first directed and guided at its very beginning."[460] Jim Langer commented: "The rise of UCSB as an internationally prominent research university was due in large part to the growing reputation of the ITP. Walter set a tone of modest, warm and thoughtful leadership especially in guiding talented young scientists at the beginnings of their careers."[460] The UCSB Chancellor Henry Yang said: "He has left a living and lasting legacy through his inspiration and impact, which is beyond anything I could adequately put into words."[460]

UCSD also published a statement noting that Walter was one of the founding faculty members in their Department of Physics. His pioneering papers with Hohenberg and Sham were mentioned, these being written when Kohn was on the faculty of UCSD. Benjamin Grinstein, Chair of the UCSD Physics Department, said: "The now famous Kohn-Sham equations... are the standard work-horse of modern materials science, and are even used in quantum theories of plasmas."[461] The University of Toronto quoted its Chemistry Nobel Laureate John Polanyi: "On winning the Nobel Prize, Walter Kohn acknowledged an enduring debt to this university, which had encouraged and inspired him at the most crucial juncture in his life."[462]

In an obituary in *Physics Today*, the magazine of the American Physical Society, when referring to the work of Walter and his collaborators, Hohenberg and Langer stated: "The reason for the enormous impact of their work is that it leads to useful approximate solutions of the many-body Schrödinger equation with no more difficulty than the Hartree–Fock method but with significantly better accuracy. As a result, DFT has been employed in a huge number of calculations of the electronic structure of solids, and since the 1980s it has been used by chemists to predict the properties of atoms and molecules of almost arbitrary complexity."[463] Hohenberg and Langer also wrote biographical memoirs of Walter Kohn for the National Academy of Sciences and for the Royal Society of London.[2,380] They described Walter as "A giant of theoretical physics… Walter's life epitomized both the hardships and wondrous achievements of physicists in the 20th century." They emphasised the vital roles of his first wife Lois and then Mara "for playing such a central role in the last decades of Walter's life". Mara died on 17 December 2018 at the age of 92 in Santa Barbara. Her collection of unique photographs taken by her father Roman Vishniac was left to the Magnes Collection of Jewish Art and Life at the University of California, Berkeley.[464]

In many interviews, Walter always emphasised the kindness of people who went out of their way to help him through his early life. Without their support he would never have gone on to make his important discoveries. In particular, he mentioned his physics teacher Emil Nohel in Vienna, the Hauff family in Sussex, Thomas Scott, the headmaster of East Grinstead School, Fritz Rothberger, who taught him the fundamentals of mathematics in the Ripples Internment Camp, the Mendel family in Toronto and Samuel Beatty who enabled him to join the University of Toronto.[23] They are all key players in his story.

Walter Kohn had an extraordinary life marked by tumultuous events and remarkable achievements. As a teenager, he witnessed the Nazi occupation of Vienna during the Anschluss in 1938 and endured arrest by the Gestapo during Kristallnacht. Fleeing to England on one of the last Kindertransport trains from Vienna in August 1939, he tragically never reunited with his parents. In 1940, labeled an "enemy alien", he was transported to Canada amidst the dangers of submarine attack. There he endured internment in a remote forest camp, many miles from civilisation. During this period, he somehow managed to study mathematics and physics. His journey led him to the universities of Toronto and Harvard, and then he established himself as a highly accomplished professor of theoretical physics in Pittsburgh,

San Diego and finally Santa Barbara. His pioneering work on DFT revolutionised computer simulations in molecular and materials science. This won him the Nobel Prize for Chemistry, a rare honour for a theoretical physicist. Beyond his scientific contributions, he fervently advocated for numerous causes, including nuclear disarmament. Walter Kohn's life stands as a testament to resilience, intellect, and unwavering dedication to progress.

References

WKP refers to the Walter Kohn papers, UArch FacP 34. Courtesy of Department of Special Research Collections, UC Santa Barbara Library, University of California, Santa Barbara.

JPP refers to Box 1/14b Job Offers, Papers of Sir John A. Pople, Science History Institute, Philadelphia.

All web pages referenced were available on 20 July 2024.

Chapter One: Vienna

1. Walter Kohn, *Nobel Voices Video History Project, 2000–2001*, Interviewer N. Hollander, 2 August 2000, Smithsonian National Museum of American History.
2. P. C. Hohenberg and J. S. Langer, Walter Kohn 9 March 1923–19 April 2016, *Biogr. Mems Fell. R. Soc.*, **64**, 249 (2018).
3. Vienna History, https://www.geschichtewiki.wien.gv.at/Salomon_Kohn.
4. Gittel Kohn (Rapaport), in https://www.geni.com/home.
5. T. Frängsmyr (Ed), *Walter Kohn, Les Prix Nobel, The Nobel Prizes 1998* (Nobel Foundation, Stockholm, 1999).
6. Walter Kohn Testimony 2012, Vancouver Holocaust Education Centre.
7. Minna Pixner (Kohn), in https://www.geni.com/home.
8. S. Beller, *Vienna and the Jews 1867–1938: A Cultural History* (Cambridge University Press, 1989).
9. Kohn brothers (Eds), *Wien* (B.K.W.1. Brüder Kohn, Wien, 1916).
10. U. Storch, *Viennese Artist Postcards from the Brüder Kohn Publishing House*, Wien Museum Magazin (26 July 2023).
11. W. Kohn, in *Walter Kohn: Personal Stories and Anecdotes Told by Friends and Collaborators*, M. Scheffler, P. Weinberger (Eds), p. 293–306 (Springer-Verlag, Berlin, 2003).
12. H. Von Hofmannsthal, *1553–1953, Vierhundert Jahre, Akademisches Gymnasium, Festschrift* (Akademisches Gymnasium, Wien, 1953).

13. R. L. Sime, *Lise Meitner: A Life in Physics* (University of California Press, Berkeley, 1996).

14. S. Friedländer, *Nazi Germany and the Jews Volume 1: The Years of Persecution, 1933–1939* (HarperCollins, New York, 1997).

15. I. Kershaw, *Hitler: A Biography* (W. W. Norton & Co., New York, 2008).

16. G. Schneider, *Exile and Destruction: The Fate of Austrian Jews 1938–1945* (Praeger, Westport CT, 1995).

17. B. Shimron, *Das Chajesrealgymnasium in Wien, 1919–1938* (Tel Aviv 5759, 1989).

18. F. Huttrer, Chajesgymnasium reunion, September 1994, *J. Assoc. Jew. Refugees Info.*, **L**, 15 (1995).

19. A. Pais, *Subtle is the Lord: The Science and the Life of Albert Einstein* (Oxford University Press, 2005).

20. D. C. Clary, *The Lost Scientists of World War II* (World Scientific Publishing, Singapore, 2024).

21. M. D. Gordin, *Einstein in Bohemia* (Princeton University Press, 2020).

22. E. Nohel, Zur natürlichen geometrie ebener transformationsgruppen (On the natural geometry of plane transformation groups), *Sitzungs-Berichte der Akademie der Wissenschaften Wien*, Bd. **123**, 2085 (1914).

23. W. Kohn, *We Would Be Separated, Quite Likely Forever*, Interview of Walter Kohn, 28 April 2014, University of California, Santa Barbara.

24. Interviews with Scientists. Interview of Walter Kohn by Tony Cheetham and John Perdew, April 2002, Film by the Vega Science Trust, http://vega.org.uk/video/programme/134.

25. H. Neuhaus, in *Walter Kohn: Personal Stories and Anecdotes Told by Friends and Collaborators*, M. Scheffler, P. Weinberger (Eds), p. 173–176 (Springer-Verlag, Berlin, 2003).

26. M. Gilbert, *Kristallnacht: Prelude to Destruction* (Harper, New York, 2007).

Chapter Two: Kindertransport

27. P. Weindling, The Kindertransport from Vienna: the children who came and those left behind, *Jewish Hist. Stud.*, **51**, 16 (2020).

28. D. Rabinovici, *Eichmann's Jews: The Jewish Administration of Holocaust Vienna, 1938–1945*, N. Somers (Trans.) (Polity, Cambridge, 2011).

29. W. Kohn, Jewish Emigrant Applications, Austria, Vienna, 1938–1939, Central Archives for the History of the Jewish People, The National Library of Israel, https://www.myheritage.com/.

30. P. Neville, *Hitler and Appeasement: The British Attempt to Prevent the Second World War* (Hambledon, London, 2006).

31. J. Bowle, *Viscount Samuel: A Biography* (Gollancz, London, 1957).

32. J. A. Cross, *Sir Samuel Hoare: A Political Biography* (Jonathan Cape, London, 1977).

33. J. Craig-Norton, *The Kindertransport: Contesting Memory* (Indiana University Press, 2019).

34. A. Grenville, The Kindertransports: An Introduction, in *The Kindertransport to Britain 1938/39: New Perspectives,* A. Hammel, B. Lewkowicz (Eds) (Rodopi, Amsterdam, 2012).

35. M. Levy, *Get the Children Out! Unsung Heroes of the Kindertransport* (Lemon Soul, London, 2023).

36. R. Holmes, The politics of compassion: the Refugee Children's Movement and caring for the Kinder, *Jewish Hist. Stud.,* **51**, 51 (2020).

37. A. Byers, *Saving Children from the Holocaust: The Kindertransport* (Leo Paper Group, Guangdong, 2012).

38. British Picture Framemakers 1600–1950, National Portrait Gallery, https://www.npg.org.uk/collections/research/programmes/conservation/directory-of-british-framemakers/h/.

39. Border Cottage, Copthorne, Borough Cuckfield, Registration district 46/1, UK Census 1939.

40. M. Kohn, Jewish Emigrant Applications, Austria, Vienna, 1938–1939, Central Library for the History of the Jewish People, The National Library of Israel, https://www.myheritage.com/.

41. L. Rosenhek, Bruch/Stücke, *Illustrierte Neue Welt*, p. 13, Ausgabe 3, 2021.

42. Minna Kohn, Female Enemy Alien Exemption From Internment Report, HO/396/47/446, National Archives, Kew.

43. Archive of the Israelitische Kultusgemeinde (IKG) Vienna, https://archiv-ikg-wien.at/media/1277/1964-1_kindertransporte-1938-1939.jpg.

44. Walter Kohn, German Jewish Aid Committee Card, A17961, Archives of World Jewish Relief.

45. D. V. Silverthorne, Vienna-London passage to safety: the portrait photographer as secondary witness in post-Anschluss émigré narratives, *Austrian Studies,* **26**, 252 (2018).

46. J. Eisinger, *Flight and Refuge: Reminiscences of a Motley Youth* (Josef Eisinger, New York, 2016).

47. East Grinstead County Grammar School (1928–1970), available from Imberhorne School Library.

48. M. Gilbert, *Churchill: A Life* (Henry Holt, New York, 1991).

49. P. Gillman and L. Gillman, *Collar the Lot: How Britain Interned and Expelled its Wartime Refugees* (Quartet Books, London, 1980).

50. N. Thompson, *The History of the Department of Physics in Bristol: 1948–1988* (H. H. Wills Physics Laboratory, Bristol, 1992).

51. A. Theobald, *Dangerous Enemy Sympathizers: Canadian Internment Camp B, 1940–1945* (Goose Lane, New Brunswick, 2019).

Chapter Three: Internment

52. D. N. Shorthose, *Properties of Matter* (William Heinemann, London, 1937).

53. R. G. Mitton, *Heat* (Dent, London, 1939).

54. A. Zangwill, The education of Walter Kohn and the creation of density functional theory, *Arch. Hist. Exact Sci.*, **68**, 775 (2014).

55. Ex-Canadian-Internees Newsletter, Number 2, February 1997.

56. C. Whitehouse, *"You'll Get Used to It!": The Internment of Jewish Refugees in Canada, 1940–43*, Ph.D. thesis, Carleton University, Ottawa (2016).

57. K. Guggenheimer, Remarques sur la constitution des noyaux atomiques. I. (Remarks on the constitution of atomic nuclei. I), *J. Phys. Rad.*, **5**, 253 (1934).

58. K. Guggenheimer, Remarques sur la constitution des noyaux-II (Remarks on the constitution of nuclei-II), *J. Phys. Rad.*, **5**, 475 (1934).

59. Guggenheimer, Dr. Kurt Martin (1902–), File 1934–48, MS. 214/5. Archives of the Society for the Protection of Science and Learning, Special Collections, Oxford University Library Services.

60. K. M. Guggenheimer, On nuclear energy levels, *Proc. R. Soc. A*, **181**, 169 (1942).

61. Rothberger, Dr. Friedrich (Fritz) (1902–), File 1939–48, MS. 284/2. Archives of the Society for the Protection of Science and Learning, Special Collections, Oxford University Library Services.

62. F. Rothberger, Sur les familles indenombrables de suites de nombres naturels et les problemes concernant la propriete C (On the innumerable families of suites of natural numbers and the problems concerning the property C), *Math. Proc. Camb. Philos. Soc.*, **37**, 109 (1941).

63. J. Lambek, Fritz Rothberger (1902–2000), *CMS Notes de la SMC*, **32**, 29, 2000.

64. A. Strutz, Forced to Flee and Deemed Suspect. Tracing Life Stories of Interned Refugees in Canada During and After the Second World War, in *Internment*

Refugee Camps, G. Anderl, L. Erker, C. Reinprecht (Eds) (Verlag, Bielefeld, 2022).

65. J. C. Slater, *Introduction to Chemical Physics* (McGraw-Hill, New York, 1939).

66. G. H. Hardy, *A Course of Pure Mathematics* (Cambridge University Press, 1921).

67. G. Ferry, *Max Perutz and the Secret of Life* (Chatto and Windus, London, 2007).

68. Fuchs, Dr. Emil Klaus Julius (1911–1988), File 1937–51, MS. 328/2. Archives of the Society for the Protection of Science and Learning, Special Collections, Oxford University Library Services.

69. N. T. Greenspan, *Atomic Spy: The Dark Lives of Klaus Fuchs* (Viking, New York City, 2020).

70. W. S. Feldberg, Bruno Mendel 1897–1959, *Biogr. Mems Fell. R. Soc.*, **6**, 191 (1960).

71. J. Eisinger, *Glimpses: A Sundry Life* (Josef Eisinger, 2023).

72. D. C. Clary, *Schrödinger in Oxford* (World Scientific Publishing, Singapore, 2022).

Chapter Four: Toronto

73. P. G. Bergmann, Leopold Infeld, authority on field theory and relativity, *Physics Today*, **21**, 113 (1968).

74. F. B. Kenrick and R. E. De Lury, *An Elementary Laboratory Course in Chemistry* (Morang, Toronto, 1905).

75. J. A. Green, Richard Dagobert Brauer, *Bull. London Math. Soc.*, **10**, 317 (1978).

76. S. Roberts and A. I. Weiss, Harold Scott MacDonald Coxeter 9 February 1907–31 March 2003, *Biogr. Mems Fell. R. Soc.*, **52**, 45 (2006).

77. R. Siegmund-Schultze, *Mathematicians Fleeing from Nazi Germany: Individual Fates and Global Impact* (Princeton University Press, 2009).

78. A. S. Eve, Sir John Cunningham McLennan, 1867–1935, *Biogr. Mems Fell. R. Soc.*, **1**, 576 (1935).

79. E. J. Allin, *Physics at the University of Toronto 1843–1980* (Department of Physics, University of Toronto, 1981).

80. D. C. Clary, Foreign Membership of the Royal Society: Schrödinger and Heisenberg?, *Notes Rec.*, **77**, 513 (2023).

81. Class Notes, Box 22, WKP.

82. J. L. Synge and B. A. Griffith, *Principles of Mechanics* (McGraw Hill, New York, 1942).

83. Weinstein, Dr. Alexander (1897–), File 1934–47, MS. 286/1. Archives of the Society for the Protection of Science and Learning, Special Collections, Oxford University Library Services.

84. A. Weinstein, On the symmetries of the solutions of a certain variational problem, *Math. Proc. Camb. Philos. Soc.*, **32**, 96 (1936).

85. W. Kohn, The spherical gyrocompass, *Quart. Appl. Math.*, **3**, 87 (1945).

86. W. Kohn, Contour integration in the theory of the spherical pendulum and the heavy symmetrical top, *Trans. Am. Math. Soc.*, **59**, 107 (1946).

87. W. Kohn, Contour integration in the theory of the spherical pendulum and the heavy symmetrical top, *Bull. Am. Math. Soc.*, **51**, 65 (1945).

88. A. F. Stevenson, A generalization of the equations of the self-consistent field for two-electron configurations, *Proc. R. Soc. A*, **160**, 588 (1937).

89. A. F. Stevenson and M. F. Crawford, A lower limit for the theoretical energy of the normal state of helium, *Phys. Rev.*, **54**, 375 (1938).

90. W. Kohn, Two applications of the variational method to quantum mechanics, *Phys. Rev.*, **71**, 635 (1947).

91. S. Lee, Sir Rudolf Ernst Peierls 5 June 1907–19 September 1995, *Biogr. Mems Fell. R. Soc.*, **53**, 265 (2007).

92. Lois M. Kohn, 22 January 2010, *The San Diego Union-Tribune*.

Chapter Five: Family and Schoolfriends

93. Franz Pixner, Universität Wien Monuments, https://monuments.univie.ac.at/index.php?title=Franz_Pixner.

94. J. Fricker, Article on Raphael Rosenhek, 14 December 2013, *Euro Echo Imaging Congress News*.

95. Franz Pixner, Dokumentationsarchiv des Österreichischen Widerstandes, Spanienarchiv.

96. Franz Pixner, Communists and suspected Communists, including Russian and Communist sympathisers, KV 2/3821, National Archives, Kew.

97. J. Vinzent, *Identity and Image. Refugee Artists from Nazi Germany in Britain (1933–1945)* (VDG, Weimar, 2006).

98. Proceedings of Minna Pixner, née Kohn (August 29, 1919), Vienna 2, Leopoldsgasse 51, against the German Reich, B Rep. 025–05 No. 5781/59, Berlin State Archives.

99. H. G. Adler, *Theresienstadt 1941–45: Das Antlitz einer Zwangsgemeinschaft, Geschichte Soziologie Psychologie (Theresienstadt 1941–45: The Face of a Compulsory Community, History Sociology Psychology)* (Mohr, Tübingen, 1955).

100. L. Rothkirchen, *The Jews of Bohemia and Moravia: Facing the Holocaust* (University of Nebraska Press, 2005).

101. K. Margry, 'Theresienstadt' (1944–1945): The Nazi propaganda film depicting the concentration camp as paradise, *Hist. J. of Film, Radio and Television*, **12**, 145 (1992).

102. A. Gottwaldt and D. Schulle, *Die 'Judendeportationen' aus dem Deutschen Reich von 1941–1945 (The 'Jewish Deportations' from the German Reich 1941–1945)* (Marix Verlag, Wiesbaden, 2005).

103. Salomon Kohn and Gittel (Gusta) Kohn, Transportation Cards, Theresienstadt to Auschwitz 28/10/1944, Doc. IDs. 5002231 and 5001007, ITS Digital Archive, Arolsen Archives.

104. Ada Levy, Eyewitness account, deportations of Jews during the Holocaust, stories of the last deportees, June 1944–April 1945, Yad Vashem Collections, https://yadvashem.org/exhibitions/last-deportees/terezin-auschwitz.html.

105. Alfred Kohn, List of persons transported from Vienna to the Łódź Ghetto, RG-15.083M.0203.00000962, United States Holocaust Memorial Museum, Copyright: Naczelna Dyrekcja Archiwów Państwowych.

106. Georg Kohn, List of persons transported from Vienna to the Łódź Ghetto, RG-15.083M.0203.00000962, United States Holocaust Memorial Museum, Copyright: Naczelna Dyrekcja Archiwów Państwowych.

107. M. Unger (Ed), *The Last Ghetto: Life in the Łódź Ghetto 1940–1944* (Yad Vashem, Jerusalem, 1995).

108. Georg, Lilly (Lily) and Hilde Kohn (Hilda Kon), Łódź, Poland, Transports to Chelmno (Kulmhof) Camp, 23 June 1944, https://www.ancestry.co.uk/.

109. Georg Kohn, Documents related to the identification and registration of Jewish Workers including identification cards, RG-15.083M.0683.00000257, United States Holocaust Memorial Museum, Copyright: Naczelna Dyrekcja Archiwów Państwowych.

110. Institut Terezínské iniciativy, https://www.holocaust.cz/en/database-of-victims/. See also https://www.doew.at/.

111. M. Raggam-Blesch, The Fate of 'Protected' Groups during the Last Years of the War. Deportations from Vienna's Nordbahnhof — a Largely Unknown Site of the Shoah, From *Deportations in the Nazi Era*, H. Borggräfe, A. Jah (Eds), Vol. 2, Arolsen Research Series (De Gruyter, Oldenbourg, 2023).

112. Herbert Neuhaus, Yad Vashem Collections, https://collections.yadvashem.org/en/deportations/7000449.

113. J. F. Entman, https://aufrichtigs.com/01-Guestbook/index.html.

114. B. Ungar-Klein, *Schattenexistenz: Jüdische U-Boote in Wien 1938–1945 (Shadow Existence: Jewish Submarines in Vienna 1938–1945)* (btb Verlag, 2021).

115. P. Salmons, https://archives.history.ac.uk/history-in-focus/Holocaust/debate.html.

116. G. Ehrlich, in *Walter Kohn: Personal Stories and Anecdotes Told by Friends and Collaborators*, M. Scheffler, P. Weinberger (Eds), p. 60–62 (Springer-Verlag, Berlin, 2003).

117. A. Brauer and G. Ehrlich, On the irreducibility of certain polynomials, *Bull. Amer. Math. Soc.*, **52**, 844 (1946).

118. H. G. Adler, Cultural Life, in *Theresienstadt 1941–1945: The Face of a Coerced Community*, A. Loewenhaar-Blauweiss (Ed), B. Cooper (Trans.) (Cambridge University Press, 2017).

119. Institut Terezínské iniciativy, https://www.holocaust.cz/en/database-of-victims/victim/48986-elise-deiner/.

120. K. Sigmund, *"Kühler Abschied von Europa" — Wien 1938 und der Exodus der Mathematik ("A Cool Farewell to Europe": Vienna 1938 and the Exodus of Mathematics)* (ÖMG, Vienna, 2001).

121. D. Miller and D. Noble, Obituary: Otto F. Hutter (1924–2020), *Physiology News Magazine*, **121**, 45 (2021).

122. O. Hutter, Exodus from Vienna, 18th Holocaust Memorial Lecture, University of Glasgow, 2018, https://www.youtube.com/watch?v=DRIrCc3TBfg.

Chapter Six: Harvard

123. P. C. Martin and S. L. Glashow, Julian Schwinger 1918–1994, *Biogr. Mem. Natl. Acad. Sci.* (2008).

124. J. Mehra and K. A. Milton, *Climbing the Mountain: The Scientific Biography of Julian Schwinger* (Oxford University Press, 2003).

125. E. Gerjuoy, *Memories of Julian Schwinger,* 2014, arXiv:1412.1410.

126. A. Zangwill, *A Mind Over Matter: Philip Anderson and the Physics of the Very Many* (Oxford University Press, 2021).

127. J. Horgan, Profile: Philip W. Anderson: gruff guru of condensed-matter physics, *Sci. Am.*, **271**, 34 (1994).

128. C. Zandonella, A lifetime of contributions to physics: Philip Anderson, Princeton University News, 18 December 2013.

129. C. P. Enz, *No Time to be Brief: A Scientific Biography of Wolfgang Pauli* (Oxford University Press, 2002).

130. W. Kohn, Joaquin M. Luttinger 1923–1997, *Biogr. Mem. Natl. Acad. Sci.* (2014).

131. W. Kohn, Variational methods in nuclear collision problems, *Phys. Rev.*, **74**, 1763 (1948).

132. D. G. Truhlar, D. W. Schwenke and D. J. Kouri, Quantum dynamics of chemical reactions by converged algebraic variational calculations, *J. Phys. Chem.*, **94**, 7346 (1990).

133. J. Z. H. Zhang and W. H. Miller, Quantum reactive scattering via the S-matrix version of the Kohn variational principle: differential and integral cross sections for $D + H_2 \rightarrow HD + H$, *J. Chem. Phys.*, **91**, 1528 (1989).

134. S. Borowitz and W. Kohn, On the electromagnetic properties of nucleons, *Phys. Rev.*, **76**, 818 (1949).

135. W. Kohn, A variation iteration method for solving secular equations, *J. Chem. Phys.*, **17**, 670 (1949).

136. F. Seitz, *The Modern Theory of Solids* (McGraw Hill, New York, 1940).

137. R. A. Silverman and W. Kohn, On the cohesive energy of metallic lithium, *Phys. Rev.*, **80**, 912 (1950).

138. W. Kohn and N. Bloembergen, Remarks on the nuclear resonance shift in metallic lithium, *Phys. Rev.*, **80**, 913 (1950); Erratum *Phys. Rev.*, **82**, 283 (1951).

139. A critical analysis of Physics S-1b, ABC 1950–1953, Box 1, WKP.

Chapter Seven: Pittsburgh

140. A. W. Tarbell, *The Story of Carnegie Tech: Being a History of Carnegie Institute of Technology from 1900 to 1935* (Carnegie Inst. Tech, 1937).

141. C. P. Slichter, Frederick Seitz 1911–2008, *Biogr. Mem. Natl. Acad. Sci.* (2010).

142. G. Hinman and D. Rose, Edward Chester Creutz 1913–2009, *Biogr. Mem. Natl. Acad. Sci.* (2010).

143. Creutz to Kohn, 14 July 1950, ABC 1950–1953, Box 1, WKP.

144. Kohn to Bohr, 24 April 1950, ABC 1950–1953, Box 1, WKP.

145. N. Bohr to National Research Council, 27 April 1950, ABC 1950–1953, Box 1, WKP.

146. National Research Council to Kohn, 31 March 1950, ABC 1950–1953, Box 1, WKP.

147. W. Kohn and Vachaspati, A difficulty in Fröhlich's theory of superconductivity, *Phys. Rev.*, **83**, 462 (1951).

148. J. Bardeen, Electron-vibration interactions and superconductivity, *Rev. Mod. Phys.*, **23**, 261 (1951).

149. Bardeen to Kohn, 28 December 1951, ABC 1950–1953, Box 1, WKP.

150. Kohn to Bardeen, 15 January 1952, ABC 1950–1953, Box 1, WKP.

151. J. Bardeen, L. N. Cooper and J. R. Schrieffer, Theory of superconductivity, *Phys. Rev.*, **108**, 1175 (1957).

152. National Research Council to Kohn, 26 December 1951, ABC 1950–1953, Box 1, WKP.

153. Creutz to Kohn, 10 November 1951, ABC 1950–1953, Box 1, WKP.

154. Kohn to Creutz, 15 November 1951, ABC 1950–1953, Box 1, WKP.

155. W. K. Stevens, Frederic de Hoffmann, 65, Dies; Physicist and Salk Institute Chief, 7 October 1989, *The New York Times.*

156. de Hoffmann to Kohn, 3 March 1952, ABC 1950–1953, Box 1, WKP.

157. Kohn to de Hoffmann, 2 April 1952, ABC 1950–1953, Box 1, WKP.

158. Peierls to Kohn, 15 November 1951, ABC 1950–1953, Box 1, WKP.

159. W. Kohn, The validity of Born expansions, *Phys. Rev.*, **87**, 539 (1952).

160. D. C. Clary and J. N. L. Connor, The vibrationally adiabatic distorted wave method for direct chemical reactions: Application to $X+F_2(v = 0, j = 0) \rightarrow XF(v', j', m_{j'})+F$, $(X = Mu, H, D, T)$, *J. Chem. Phys.*, **75**, 3329 (1981).

161. Beatty to Kohn, 27 September 1951, ABC 1950–1953, Box 1, WKP.

162. W. Kohn, D. Ruelle and A. Wightman, Obituary: Res Jost, *Physics Today*, **45**, 120 (1992).

163. R. Jost and W. Kohn, Construction of a potential from a phase shift, *Phys. Rev.*, **87**, 977 (1952).

164. Kohn to Beatty, Hand-written letter 1951, ABC 1950–1953, Box 1, WKP.

165. Creutz to Kohn, 22 May 1952, ABC 1950–1953, Box 1, WKP.

166. M. Kohn to W. Kohn, Handwritten letter July 1952, ABC 1950–1953, Box 1, WKP.

167. Kohn to Borowitz, 21 October 1952, ABC 1950–1953, Box 1, WKP.

168. Kohn to Krieger, 17 November 1952, KLMNOP 1952–1953, Box 1, WKP.

169. Kohn to Coxeter, 28 November 1952, ABC 1950–1953, Box 1, WKP.

170. Coxeter to Kohn, 17 December 1952, ABC 1950–1953, Box 1, WKP.

171. Herzberg to Kohn, 16 December 1952, GHIJ 1952–1953, Box 1, WKP.

172. Kohn to Luttinger, 30 December 1952, KLMNOP 1952–1953, Box 1, WKP.

173. H. Tate (Chair of Mathematics) to Kohn, 13 February 1953, QRST 1952–1953, Box 1, WKP.

174. Kohn to Wallace, 2 January 1953, UVWXYZ 1952–1953, Box 1, WKP.

175. Wallace to Kohn, 5 January 1953, UVWXYZ 1952–1953, Box 1, WKP.

176. Eisinger to Kohn, Undated, 1952, DEF 1952–1953, Box 1, WKP.

177. R. A. Deller (Bell Labs) to Kohn, 12 January 1953, ABC 1950–1953, Box 1, WKP.

178. Creutz to Kohn, 24 February 1953, ABC 1950–1953, Box 1, WKP.

179. Bell Labs to Kohn, 21 May 1953, ABC 1950–1953, Box 1, WKP.

180. W. Kohn and N. Rostoker, Solution of the Schrödinger equation in periodic lattices with an application to metallic lithium, *Phys. Rev.*, **94**, 1111 (1954).

181. R. C. Fletcher (Bell Labs) to Kohn, 12 November 1953, DEF 1953–1954, Box 1, WKP.

182. Bell Labs to Kohn, 19 May 1954, ABC 1953–1954, Box 1, WKP.

183. J. M. Luttinger and W. Kohn, Motion of electrons and holes in perturbed periodic fields, *Phys. Rev.*, **97**, 869 (1955).

184. Sondhoff to Kohn, 16 June 1953, QRST 1959–1960, Box 1, WKP.

185. Ernest Sterzer Memoir, United States Holocaust Memorial Museum.

186. Seitz to Kohn, 1 February 1955, QRST 1954–1955, Box 1, WKP.

187. S. Kaya to Kohn, 19 July 1954, GHIJ 1953–1954, Box 1, WKP.

188. Rotterdamsche Bank to Kohn, 13 May 1954, QRST 1953–1954, Box 1, WKP.

189. Kohn to Jost, 24 November 1954, GHIJ 1954–1955, Box 1, WKP.

190. Kohn to Busch, 9 May 1955, ABC 1954–1955, Box 1, WKP.

191. Kohn to Koenig, 19 May 1955, DEF 1954–1955, Box 1, WKP.

192. Eisinger to Kohn, 1 November 1954, DEF 1954–1955, Box 1, WKP.

193. J. D. Martin, *Solid State Insurrection: How the Science of Substance Made American Physics Matter* (University of Pittsburgh Press, 2018).

194. Jones to Kohn, 17 May 1955, GHIJ 1954–1955, Box 1, WKP.

195. K. A. Brueckner, Many-body problem for strongly interacting particles. II. Linked cluster expansion, *Phys. Rev.*, **100**, 36 (1955).

196. K. A. Brueckner and C. A. Levinson, Approximate reduction of the many-body problem for strongly interacting particles to a problem of self-consistent fields, *Phys. Rev.*, **97**, 1344 (1955).

197. M. Gell-Mann and K. A. Brueckner, Correlation energy of an electron gas at high density, *Phys. Rev.*, **106**, 364 (1957).

198. W. Kohn, Effective mass theory in solids from a many-particle standpoint, *Phys. Rev.*, **105**, 509 (1957).

199. Warner to Kohn, 29 January 1958, UVWXYZ 1957–1958, Box 2, WKP.

200. R Pariser and R. G. Parr, A semi-empirical theory of the electronic spectra and electronic structure of complex unsaturated molecules. I, *J. Chem. Phys.*, **21**, 466 (1953).

201. J. A. Pople, Electron interaction in unsaturated hydrocarbons, *Trans. Faraday Soc.*, **49**, 1375 (1953).

202. Herring to Kohn, 6 January 1959, GHIJ 1958–1959, Box 2, WKP.

203. P. W. Anderson, in *Walter Kohn: Personal Stories and Anecdotes Told by Friends and Collaborators*, M. Scheffler, P. Weinberger (Eds), p. 3–5 (Springer-Verlag, Berlin, 2003).

204. W. Kohn, Image of the Fermi surface in the vibration spectrum of a metal, *Phys. Rev. Lett.*, **2**, 393 (1959).

205. S. Piscanec, M. Lazzeri, F. Mauri, A. C. Ferrari and J. Robertson, Kohn anomalies and electron-phonon interactions in graphite, *Phys. Rev. Lett.*, **93**, 185503 (2004).

206. W. Kohn and S. H. Vosko, Theory of nuclear resonance intensity in dilute alloys, *Phys. Rev.*, **119**, 912 (1960).

Chapter Eight: San Diego

207. T. F. Malone, E. D. Goldberg and W. H. Munk, Roger Randall Dougan Revelle 1909–1991, *Biogr. Mem. Natl. Acad. Sci.* (1998).

208. M. McClain, The Scripps family's San Diego experiment, *J. San Diego Hist.*, **56**, 1 (2010).

209. *A Master Plan for Higher Education in California 1960–1975* (California State Department of Education, Sacramento, 1960).

210. C. Kerr, *The Gold and the Blue: A Personal Memoir of the University of California, 1949–1997. Academic Triumphs* (University of California Press, 2001).

211. M. H. Thiemens, James R. Arnold: From the Manhattan Project to the moon and beyond, *Proc. Natl. Acad. Sci.*, **109**, 4339 (2012).

212. J. R. Arnold, J. Bigeleisen and C. A. Hutchison Jr., Harold Clayton Urey 1893–1981, *Biogr. Mem. Natl. Acad. Sci.* (1995).

213. R. C. Fahey, C. Kohl, K. Marti and M. Thiemens, In Memoriam. James R. Arnold, Professor Emeritus of Chemistry and Biochemistry, UC San Diego, 1926–2012 (University of California, 2012).

214. W. Bialek, Keith Brueckner March 19, 1924–September 19, 2014, *Biogr. Mem. Natl. Acad. Sci.* (2023).

215. N. S. Anderson, *An Improbable Venture: A History of the University of California, San Diego* (UCSD Press, La Jolla, 1993).

216. Revelle to Kohn, 29 May 1959, La Jolla — General 1959–1960, Box 3, WKP.

217. Revelle to Kohn, 19 July 1959, La Jolla — General 1959–1960, Box 3, WKP.

218. Kohn to Warner, 22 October 1959, KLM 1959–1960, Box 2, WKP.

219. Warner to Kohn, 4 November 1959, UVWXYZ 1959–1960, Box 2, WKP.

220. Creutz to Kohn, 25 February 1960, ABC 1959–1960, Box 2, WKP.

221. Brueckner to Kohn, 17 September 1959, ABC 1959–1960, Box 2, WKP.

222. Kohn to Mottelson, 26 October 1959, KLM 1959–1960, Box 2, WKP.

223. R. Revelle, Interviewed by K. Ringrose, May 15–16 1985, University of California, San Diego, 25th Anniversary Oral History Project.

224. P. W. Anderson, Interviewed by A. Kojevnikov, 23 November 1999, Niels Bohr Library and Archives, American Institute of Physics, College Park, MD USA.

225. L. Peterson, Interviewed by D. Zierler, 4 November 2020, Niels Bohr Library and Archives, American Institute of Physics, College Park, MD USA.

226. Condon to Kohn, 11 November 1959, ABC 1959–1960, Box 2, WKP.

227. Kohn to Condon, 20 November 1959, ABC 1959–1960, Box 2, WKP.

228. Brueckner to Kohn, 1 December 1959, ABC 1959–1960, Box 2, WKP.

229. M. E. Stratthaus, Flaw in the jewel: housing discrimination against Jews in La Jolla, California, *Am. Jewish Hist.,* **84**, 189 (1996).

230. R. G. Sachs, Maria Goeppert Mayer June 28, 1906–February 20, 1972, *Biogr. Mem. Natl. Acad. Sci.* (1979).

231. A. I. Sargent and M. S. Longair, Eleanor Margaret Burbidge 12 August 1919–5 April 2020, *Biog. Mems. Fell. R. Soc.,* **71**, 11 (2021).

232. Program in Physics, Senate Business — Extension pt.1 1960–1961, Box 3, WKP.

233. La Jolla Physics Symposium, September 6–8 1985, University of California, San Diego.

234. M. H. Cohen (U. of Chicago) to Kohn, 8 March 1960, ABC 1959–1960, Box 2, WKP.

235. Wigner to Kohn, 8 March 1960, UVWXYZ 1959–1960, Box 2, WKP.

236. Kohn to Brueckner, 26 October 1959, ABC 1959–1960, Box 2, WKP.

237. Lily Rapaport to Kohn, 1 January 1960, QRST 1959–1960, Box 2, WKP.

238. Kohn to Jost, 17 February 1960, GHIJ 1958–1959, Box 2, WKP.

239. Oliver E. Buckley Condensed Matter Physics Prize, *Prizes and Awards*, American Physical Society.

240. W. Kohn and S. J. Nettel, Giant fluctuations in a degenerate Fermi gas, *Phys. Rev. Lett.,* **5**, 8 (1960).

241. Kohn to Brown, June 1960, ABC 60–61, Box 3, WKP.

242. Brown to Kohn, 24 June 1960, ABC 60–61, Box 3, WKP.

243. Salk to Kohn, 22 March 1960, QRST 1959–1960, Box 2, WKP.

244. Kerr to Kohn, 24 June 1963, Professional Correspondence 1964–1966, Box 4, WKP.

245. W. Kohn, Theory of the insulating state, *Phys. Rev. A,* **133**, 171 (1964).

246. Wigner to Kohn, 29 January 1964, Professional Correspondence 1964–1966, Box 4, WKP.

247. P. Nozières and J. M. Luttinger, Derivation of the Landau Theory of Fermi liquids. I. Formal preliminaries, *Phys. Rev.,* **127**, 1423 (1962).

248. P. C. Hohenberg and P. C. Martin, Superfluid dynamics in the hydrodynamic and collisionless domains, *Phys. Rev. Lett.,* **12**, 69 (1964).

249. P. Hohenberg, Anisotropic superconductors with nonmagnetic impurities, *Sov. Phys. JETP-USSR,* **18**, 834 (1964).

250. Walter Kohn, Biographical, The Nobel Prize for Chemistry 1998, NobelPrize. org, Nobel Prize Outreach.

251. P. Hohenberg and W. Kohn, Inhomogeneous electron gas, *Phys. Rev.,* **136**, B864 (1964).

252. N. F. Mott and H. S. W. Massey, *The Theory of Atomic Collisions* (Clarendon Press, Oxford, 1933).

253. N. F. Mott, Introduction, *Proc. R. Soc. A,* **371**, 3 (1980).

254. Pippard to Kohn, 7 November 1963, Professional Correspondence 1964–1966, Box 4, WKP.

255. Trainor to Kohn, 29 October 1963, Professional Correspondence 1964–1966, Box 4, WKP.

256. Kohn to Hohenberg, 15 June 1964, Professional Correspondence 1964–1966, Box 4, WKP.

257. Kohn to R. E. Prange (University of Maryland), 3 March 1964, Professional Correspondence 1964–1966, Box 4, WKP.

258. L. J. Sham, Interviewed by D. Zierler, 22 October 2020, Niels Bohr Library and Archives, American Institute of Physics, College Park, MD USA.

259. L. J. Sham, A calculation of the phonon frequencies in sodium, *Proc. R. Soc.,* **283**, 33 (1965).

260. Kohn to Mermin, 13 October 1964, Many-Body Problem Conference, Box 5, WKP.

261. W. Kohn and L. J. Sham, Self-consistent equations including exchange and correlation effects, *Phys. Rev.,* **140**, A1133 (1965).

262. D. M. Ceperley and B. J. Alder, Ground state of the electron gas by a stochastic method, *Phys. Rev. Lett.,* **45**, 566 (1980).

263. Hopfield to Kohn, 4 December 1963, Professional Correspondence 1964–1966, Box 4, WKP.

264. B. Y. Tong and L. J. Sham, Application of a self-consistent scheme including exchange and correlation effects to atoms, *Phys. Rev.,* **144**, 1 (1966).

265. N. D. Mermin, *Why Quark Rhymes with Pork: and Other Scientific Diversions* (Cambridge University Press, 2016).

266. N. D. Mermin, in *Walter Kohn: Personal Stories and Anecdotes Told by Friends and Collaborators*, M. Scheffler, P. Weinberger (Eds), p. 155–159 (Springer-Verlag, Berlin, 2003).

267. N. D. Mermin, Thermal properties of the inhomogeneous electron gas, *Phys. Rev. A*, **137**, 1441 (1965).

268. Kohn to Mermin, 28 July 1964, Many-Body Problem Conference, Box 5, WKP.

269. Kohn to Celli, 13 January 1965, UCSD — Galbraith Resignation 1965–1966, Box 5, WKP.

270. Harvard: Inhomogeneous Electron Gas 1966, Box 37, WKP.

271. W. Kohn, A New Formulation of the Inhomogeneous Electron Gas Problem, in *Many Body Theory*, R. Kubo (Ed) (W.A. Benjamin, New York 1966).

272. Correspondence, UCSD — Galbraith Resignation 1965–1966, Box 5, WKP.

273. W. Kohn, Mott and Wigner transitions, *Phys. Rev. Lett.,* **19**, 789 (1967).

274. Mott to Kohn, 3 October 1967, UCSD — Galbraith Resignation 1965–1966, Box 5, WKP.

275. N. F. Mott and E. A. Davis, Conduction in non-crystalline systems. II. The metal-insulator transition in a random array of centres, *Philos. Mag.,* **17**, 1269 (1968).

276. D. Jérome, T. M. Rice and W. Kohn, Excitonic insulator, *Phys. Rev.,* **158**, 462 (1967).

277. Bissell to Kohn, 9 March 1967, La Jolla — General 1959–1960, Box 3, WKP.

278. N. D. Lang and W. Kohn, Theory of metal surfaces: charge density and surface energy, *Phys. Rev. B*, **1**, 4555 (1970).

279. N. D. Lang and W. Kohn, Theory of metal surfaces: work function, *Phys. Rev. B*, **3**, 1215 (1971).

280. Davisson-Germer Prize in Atomic or Surface Physics, *Prizes and Awards*, American Physical Society.

281. W. Jones and W. H. Young, Density functional theory and the von Weizsacker method, *J. Phys. C: Solid State Phys.,* **4**, 1322 (1971).

282. J. J. Rehr, in *Walter Kohn: Personal Stories and Anecdotes Told by Friends and Collaborators*, M. Scheffler, P. Weinberger (Eds), p. 198–199 (Springer-Verlag, Berlin, 2003).

Chapter Nine: DFT

283. R. G. Parr in *Chemical Reactivity Theory: A Density Functional View*, P. K. Chattaraj (Ed) (CRC Press, Boca Raton, 2009).

284. W. Kohn, in A Celebration of the Contributions of Robert G. Parr, *Reviews of Modern Quantum Chemistry*, K. D. Sen (Ed), Vol. 1, v (World Scientific Publishing, Singapore, 2002).

285. E. S. Kryachko, Hohenberg-Kohn theorem, *Int. J. Quant. Chem.*, **18**, 1029 (1980).

286. M. Levy and J. P. Perdew, In defense of the Hohenberg-Kohn Theorem and density functional theory, *Int. J. Quant. Chem.*, **21**, 511 (1982).

287. W. Heitler and F. London, Wechselwirkung neutraler atome und homöopolare bindung nach der quantenmechanik (Interaction of neutral atoms and homopolar bonding according to quantum mechanics), *Z. Phys.*, **44**, 455 (1927).

288. J. E. Lennard-Jones, The electronic structure of some diatomic molecules, *Trans. Faraday. Soc.*, **25**, 668 (1929).

289. C. C. J. Roothaan, Self-Consistent Field theory for open shells of electronic systems, *Rev. Mod. Phys.*, **32**, 179 (1960).

290. E. A. Hylleraas, Neue berechnung der energie des heliums im grundzustande, sowie des tiefsten terms von ortho-helium (New calculation of the energy of helium in the ground state, as well as the lowest term of ortho-helium), *Z. Phys.*, **54**, 347 (1929).

291. D. C. Clary, Variational calculations on many-electron diatomic molecules using Hylleraas-type wavefunctions, *Mol. Phys.*, **34**, 793 (1977).

292. C. J. Cramer, *Essentials of Computational Chemistry: Theory and Models* (Wiley, Chichester, 2002).

293. L. H. Thomas, The calculation of atomic fields, *Math. Proc. Camb. Philos. Soc.*, **23**, 542 (1927).

294. E. Fermi, Un metodo statistico per la determinazione di alcune priopprietà dell'atomo (A statistical method for determining some properties of the atom), *Rend. Lincei*, **6**, 602 (1927).

295. P. A. M. Dirac, Note on exchange phenomena in the Thomas atom, *Math. Proc. Camb. Philos. Soc.*, **26**, 376 (1930).

296. E. Teller, On the stability of molecules in the Thomas-Fermi theory, *Rev. Mod. Phys.*, **34**, 627 (1962).

297. J. W. D. Connolly, QTP in the 60s: John C. Slater and the beginnings of density functional theory, *Mol. Phys.*, **108**, 2863 (2010).

298. R. G. Parr, in *Walter Kohn: Personal Stories and Anecdotes Told by Friends and Collaborators*, M. Scheffler, P. Weinberger (Eds), p. 187–191 (Springer-Verlag, Berlin, 2003).

299. E. B. Wilson, Four-dimensional electron density function, *J. Chem. Phys.*, **36**, 2232 (1962).

300. M. Levy, Universal variational functionals of electron densities, first-order density matrices, and natural spin-orbitals and solution of the v-representability problem, *Proc. Natl. Acad. Sci.*, **76**, 6062 (1979).

301. E. H. Lieb and S. Oxford, Improved lower bound on the indirect Coulomb energy, *Int. J. Quant. Chem.*, **19**, 427 (1981).

302. D. C. Langreth and M. J. Mehl, Beyond the local-density approximation in calculations of ground-state electronic properties. *Phys. Rev. B*, **28**, 1809 (1983).

303. J. P. Perdew and A. Zunger, Self-interaction correction to density-functional approximations for many-electron systems, *Phys. Rev. B*, **23**, 5048 (1981).

304. J. P. Perdew, My life in science: lessons for yours?, *J. Chem. Phys.*, **160**, 010402 (2024).

305. C. Lee, W. Yang and R. G. Parr, Development of the Colle-Salvetti correlation energy formula into a functional of the electron density, *Phys. Rev. B*, **37**, 785 (1988).

306. A. D. Becke, Density-functional exchange-energy approximation with correct asymptotic behavior, *Phys. Rev. A*, **38**, 3098 (1988).

307. R. G. Parr, Nomination of Walter Kohn to IAQMS, 8 April 1991, Papers of Robert G. Parr, Science History Institute, Philadelphia.

308. M. Charlton, Highlights from Axel Becke's Killam Lecture, 7 December 2016, Dal News, Dalhousie University.

309. B. G. Johnson, P. M. W. Gill and J. A. Pople, The performance of a family of density functional methods, *J. Chem. Phys.*, **98**, 5612 (1993).

310. D. J. Tozer, Nicholas Handy and density functional theory, *Mol. Phys.*, **103**, 145 (2005).

311. G. E. Scuseria and V. N. Staroverov, Progress in the Development of Exchange-Correlation Functionals, Chapter 24 in *Theory and Applications of Computational Chemistry: The First 40 Years (A Volume of Technical and Historical Perspectives)*, C. E. Dykstra, G. Frenking, K. S. Kim, G. E. Scuseria (Eds) (Elsevier, Amsterdam, 2005).

312. R. O. Jones, Density functional theory: Its origins, rise to prominence, and future, *Rev. Mod. Phys.*, **87**, 897 (2015).

313. K. Berland, V. R. Cooper, K. Lee, E. Schröder, T. Thonhauser, P. Hyldgaard and B. I. Lundqvist, Van der Waals forces in density functional theory: a review of the vdW-DF method, *Rep. Prog. Phys.*, **78**, 066501 (2015).

314. M. Bühl, C. Reimann, D. A. Pantazis, T. Bredow and F. Neese, Geometries of third-row transition-metal complexes from density-functional theory, *J. Chem. Theory Comput.*, **4**, 1449 (2008).

315. W. Liu and Y. Xiao, Relativistic time-dependent density functional theories, *Chem. Soc. Rev.*, **47**, 4481 (2018).

316. R. Car and M. Parrinello, Unified approach for molecular dynamics and density-functional theory, *Phys. Rev. Lett.*, **55**, 2471 (1985).

317. E. Runge and E. K. U. Gross, Density-functional theory for time-dependent systems, *Phys. Rev. Lett.*, **52**, 997 (1984).

Chapter Ten: Pople

318. John Pople, Interviewed by S. Forsén, Meeting of Nobel Laureates in Lindau, Germany, June 2000, Nobel Prize Outreach, https://www.nobelprize.org/prizes/chemistry/1998/pople/interview/.

319. John Pople, Biographical, The Nobel Prize for Chemistry 1998, NobelPrize.org, Nobel Prize Outreach.

320. A. D. Buckingham, Sir John Anthony Pople KBE 31 October 1925–15 March 2004, *Biog. Mems Fell. R. Soc.*, **52**, 299 (2006).

321. N. F. Mott, John Edward Lennard-Jones 1894–1954, *Biog. Mems Fell. R. Soc.*, **1**, 174 (1955).

322. J. E. Lennard-Jones and J. A. Pople, The molecular orbital theory of chemical valency. IV. The significance of equivalent orbitals, *Proc. R. Soc. A.*, **202**, 166 (1950).

323. J. A. Pople, The molecular orbital theory of chemical valency. V. The structure of water and similar molecules, *Proc. R. Soc. A.*, **202**, 323 (1950).

324. J. A. Pople, Molecular association in liquids. II. A theory of the structure of water, *Proc. R. Soc. A*, **205**, 163 (1951).

325. Records of the Kaptiza Club, Ref. GBR/0014/CKFT 7, Papers of Sir John Cockcroft, Churchill Archives Centre, Churchill College, Cambridge.

326. U. Anders, Interview with Professor Robert G. Parr, Early Ideas in the History of Quantum Chemistry, 2001, http://quantum-chemistry-history.com/Parr1.htm.

327. R. L. Gregory and J. N. Murrell, Hugh Christopher Longuet–Higgins 11 April 1923–27 March 2004, *Biog. Mems Fell. R. Soc.*, **52**, 149 (2006).

328. J. A. Pople and R. K. Nesbet, Self-consistent orbitals for radicals, *J. Chem. Phys.*, **22**, 571 (1954).

329. J. A. Pople, W. G. Schneider and H. J. Bernstein, *High Resolution Nuclear Magnetic Resonance* (McGraw-Hill, New York, 1959).

330. H. Sisler to Pople, 29 July 1958, JPP.

331. Parr to Pople, 25 January 1963, JPP.

332. Allen to Hornig, 13 February 1963, JPP.

333. W. Kauzmann (Princeton) to Pople, 21 March 1963, JPP.

334. H. P. Smith (Bell) to Pople, 1 April 1963, JPP.

335. Hornig (Princeton) to Pople, 1 April 1963, JPP

336. Mulliken to Pople, 20 March 1963, JPP.

337. E. Levi (Chicago) to Pople, 18 March 1963, JPP.

338. Schneider to Pople, 15 March 1963, JPP.

339. R. Carlin to Pople, 31 October 1962, JPP.

340. Cross to Pople, 18 December 1962, JPP.

341. G. Gee to Pople, 2 April 1963, JPP.

342. Coulson to Pople, 26 March 1963, JPP.

343. Sutherland to Pople, 28 March 1963, JPP.

344. H. Hadow (Nat. Phys. Lab.) to Pople, 14 January 1964, JPP.

345. Newspaper clippings, Box 1/16, Papers of Sir John A. Pople, Science History Institute, Philadelphia.

346. Hilary Pople (daughter of John Pople), private communication to the author.

347. C. A. Coulson, Samuel Francis Boys 1911–1972, *Biog. Mems Fell. R. Soc.*, **19**, 94 (1973).

348. W. J. Hehre, W. A. Lathan, R. Ditchfield, M. D. Newton and J. A. Pople, Gaussian 70, Quantum Chemistry Program Exchange, Program No. 237 (1970).

349. John Pople, *Nobel Voices Video History Project, 2000–2001*, Interviewer N. Hollander, 29 June 2000, Smithsonian National Museum of American History.

350. John A. Pople, Wolf Prize Laureate in Chemistry 1992, Wolf Foundation, https://wolffund.org.il/john-a-pople/.

Chapter Eleven: Santa Barbara

351. J. S. Langer, My career as a theoretical physicist — so far, *Annu. Rev. Condens. Matter Phys.*, **8**, 1 (2017).

352. J. Hartle, W. Kohn, D. Scalapino and R. Sugar, Robert Huttenback: 1928–2012, 25 July 2012, *Santa Barbara Independent*.

353. Walter Kohn Security Clearance Application, FBI Security File, Ref. 116A-34618, July 1987, for US Department of Energy, FOI/PA No. 1349200.

354. M. Wiesel (Ed), *To Give Them Light: The Legacy of Roman Vishniac* (Simon & Schuster, New York, 1993).

355. *Vishniac*, a film by Laura Bialis (2023).

356. D. J. Scalapino and R. Sugar, Walter Kohn (1923–2016), *Proc. Natl. Acad. Sci.,* **113**, 8883 (2016).

357. Who is the Legendary Gang of Four?, University of California, Santa Barbara, https://giving.ucsb.edu/Impact/gang-of-four.

358. H. Gross, in *Walter Kohn: Personal Stories and Anecdotes Told by Friends and Collaborators*, M. Scheffler, P. Weinberger (Eds), p. 82–84 (Springer-Verlag, Berlin, 2003).

359. R. Sugar, in *Walter Kohn: Personal Stories and Anecdotes Told by Friends and Collaborators*, M. Scheffler, P. Weinberger (Eds), p. 252–256 (Springer-Verlag, Berlin, 2003).

360. Love to Kohn, 16 October 1998, Nobel 1998–2001, Box 146, WKP.

361. L. J. Sham, in *Walter Kohn: Personal Stories and Anecdotes Told by Friends and Collaborators*, M. Scheffler, P. Weinberger (Eds), p. 234–237 (Springer-Verlag, Berlin, 2003).

362. National Science and Technology Medals Foundation, https://nationalmedals.org/laureate/walter-kohn/.

363. M. H. Kalos and P. Vashishta, Presentation of the Fourth Eugene Feenberg Memorial Medal in Many-Body Physics to Walter Kohn, in *Recent Progress in Many-Body Theories*, Vol. 3, T. L. Ainsworth *et al.* (Eds), (Plenum Press, New York, 1992).

364. UNESCO Awards, https://www.unesco.org/en/prizes.

Chapter Twelve: Nobel

365. Alfred Nobel's Will, https://www.nobelprize.org/alfred-nobel/alfred-nobels-will/.

366. The Nobel Prize, https://www.nobelprize.org/about/the-nobel-committee-for-physics/.

367. Nobel Prize Case Study: Arnold Sommerfeld, Center for History of Physics, American Institute of Physics.

368. Nobel Prize Organisation, https://www.nobelprize.org/prizes/lists/all-nobel-prizes-in-physics/.

369. J. P. Perdew, in *Walter Kohn: Personal Stories and Anecdotes Told by Friends and Collaborators*, M. Scheffler, P. Weinberger (Eds), p. 194–195 (Springer-Verlag, Berlin, 2003).

370. Lundqvist to Kohn, 13 October 1998, Nobel 1998–2001, Box 146, WKP.

371. International Academy of Quantum Molecular Science, https://iaqms.org/members.php.

372. The Welch Foundation, https://welch1.org/awards/welch-award-in-chemistry/welch-award-guidelines.

373. R. G. Parr, Nomination of Walter Kohn for the Welch Award, 28 January 1998, Box 34/17, Papers of Robert G. Parr, Science History Institute, Philadelphia.

374. D. C. Clary and B. J. Orr (Eds), *Optical, Electric and Magnetic Properties of Molecules: A Review of the Work of A. D. Buckingham* (Elsevier, Amsterdam, 1997).

375. Royal Swedish Academy of Sciences, Press Release, 13 October 1998, NobelPrize.org, Nobel Prize Outreach.

376. NU Professor Shares Nobel for Chemistry, 14 October 1998, *Chicago Tribune.*

377. Five Scientists in US Get Nobels, 13 October 1998, *The New York Times.*

378. Norrby to Kohn, 13 October 1998, Nobel 1998–2001, Box 146, WKP.

379. Nobel 1998–2001, Box 146, WKP.

380. P. C. Hohenberg and J. S. Langer, Walter Kohn 1923–2016, *Biogr. Mem. Natl. Acad. Sci.* (2017).

381. J. I. Seeman, Ernest Ludwig Eliel, 1921–2008, *Biogr. Mem. Natl. Acad. Sci.* (2014).

382. R. G. Newton, *Scattering Theory of Waves and Particles* (Springer, Berlin, 1982).

383. Box 3/13b Nobel Prize — congratulatory letters, Papers of Sir John A. Pople, Science History Institute, Philadelphia.

384. W. Kohn, Nobel Lecture: Electronic structure of matter — wave functions and density functionals, *Rev. Mod. Phys., **71**,* 1253 (1999).

385. B. Roos, Nobel Prize for Chemistry 1998, Award ceremony speech, Nobel Prize Outreach, https://www.nobelprize.org/prizes/chemistry/1998/ceremony-speech/.

386. Nate D. Sanders, Auctions, Nobel Prize Won by Walter Kohn, One of the Children Saved by *Kindertransport* During World War II, January 2022.

Chapter Thirteen: Politics

387. Walter Kohn, Interview by Marika Griehsel, 2004, 54th meeting of Nobel Laureates in Lindau, Germany, Nobel Prize Outreach, https://www.nobelprize.org/prizes/chemistry/1998/kohn/interview/.

388. 11 at Carnegie Tech Ask Bomb Test Pact, 31 October 1956, *The New York Times.*

389. R. Bellah, O. Chamberlain and W. Kohn, Weapons Labs: Light-Years From the Teaching Mission, 1 August 1990, *Los Angeles Times*.

390. Statement by Nobel Laureates, The Next 100 Years, 7 December 2001, *Times Higher Education*.

391. The US Congress Should Act Against Nukes, 14 February 2007, *The New York Times*.

392. W. Kohn, in *Science and Human Rights,* C. Corillon (Ed), p. 75 (National Academies Press, Washington DC, 1988).

393. Perspectives on the Crisis of UNESCO, W. Kohn, F. Newman, R. Revelle (Eds), IGCC, Box 158, WKP.

394. W. Kohn and R. Revelle, UNESCO Director General, *Science*, **236**, 1503 (1987).

395. W. Kohn, Science and Peace, UNESCO TV (2011).

396. F. Riggs (Dept. of State) to McElroy, Professor Khalatnikov, 1974–1975, Box 6, WKP.

397. Kohn to *Prensa Popular*, 14 February 1974, Professor Khalatnikov, 1974–1975, Box 6, WKP.

398. G. Bravo, *Viktor Ullmann*, Composers, The Orel Foundation.

399. S. Baek, "A Music Offering" Echoes Sounds of the Holocaust, 27 October 2010, *The UCSB Bottom Line*.

400. E. Broclawik and K. Broclawik, in *Walter Kohn: Personal Stories and Anecdotes Told by Friends and Collaborators*, M. Scheffler, P. Weinberger (Eds), p. 26–29 (Springer-Verlag, Berlin, 2003).

401. L. Adler-Kastner, A letter from Vienna, *The Edinburgh Star*, **36**, 14 (2000).

402. P. Weinberger, in *Walter Kohn: Personal Stories and Anecdotes Told by Friends and Collaborators*, M. Scheffler, P. Weinberger (Eds), p. 267–274 (Springer-Verlag, Berlin, 2003).

403. Austria and its Decorations, https://www.bundespraesident.at/ehrenzeichen/.

404. K. Hanta, From Exile to Excellence. Interview with Nobel Prize Laureate Walter Kohn, *Austria Kultur*, Vol. 9 No. 1, January/February (1999).

405. K. Schwarz, in *Walter Kohn: Personal Stories and Anecdotes Told by Friends and Collaborators*, M. Scheffler, P. Weinberger (Eds), p. 227–230 (Springer-Verlag, Berlin, 2003).

406. Akademisches Gymnasium Wien, https://akg-wien.at/schulprojekte/.

407. E. R. Kandel, *In Search of Memory: The Emergence of a New Science of Mind* (W. W. Norton & Co., New York, 2007).

408. Honorary doctorates to Alfred Bader, Walter Kohn and Peter Pulzer, 27 November 2012, *Universität Wien Press*.

409. The Vienna Project, *Österreich J.*, **136**, 76 (2014).

410. Bachelet to Kohn, 5 February 2000, Box 146, WKP.
411. Kohn to Bachelet, 5 February 2000, Box 146, WKP.
412. Pope Makes Historic Pardon Request for Church Sins, 12 March 2000, CNN.
413. Kohn to Bachelet, 1 April 2000, Box 146, WKP.
414. G. Bachelet, in *Walter Kohn: Personal Stories and Anecdotes Told by Friends and Collaborators*, M. Scheffler, P. Weinberger (Eds), p. 15–19 (Springer-Verlag, Berlin, 2003).
415. J. Życiński, in *Walter Kohn: Personal Stories and Anecdotes Told by Friends and Collaborators*, M. Scheffler, P. Weinberger (Eds), p. 288–292 (Springer-Verlag, Berlin, 2003).
416. Kohn to Bachelet, 5 March 2001, Box 146, WKP.
417. Walter Kohn: Reflections of a Physicist after an Encounter with the Vatican and Pope John Paul II, 6 August 2001, University of California TV.
418. T. Feder, Nobelist creates films on solar power, *Physics Today,* **59**, 26 (2006).
419. The Nobel Prize for Chemistry 2000, NobelPrize.org, Nobel Prize Outreach.
420. Shuji Nakamura Wins Nobel Prize in Physics, 6 October 2014, UC Santa Barbara Materials, https://materials.ucsb.edu/news/shuji-nakamura-wins-nobel-prize-physics.

Chapter Fourteen: Legacy

421. H. A. Fertig and W. Kohn, Symmetry of the atomic electron density in Hartree, Hartree-Fock, and density-functional theories, *Phys. Rev. A,* **62**, 052511 (2000).
422. W. Kohn, A. Savin and C. A. Ullrich, Hohenberg–Kohn theory including spin magnetism and magnetic fields, *Int. J. Quant. Chem.,* **100**, 20 (2004).
423. K. Schwarz, L. J. Sham, A. E. Mattsson and M. Scheffler, Obituary for Walter Kohn (1923–2016), *Computation,* **4**, 40 (2016).
424. E. Schrödinger, Discussion of probability relations between separated systems, *Math. Proc. Camb. Philos. Soc.,* **31**, 555 (1935).
425. E. Schrödinger, Die gegenwärtige situation in der quantenmechanik (The present situation in quantum mechanics), *Naturwissenschaften,* **23**, 807 (1935).
426. Visitors Book of the President's Lodgings (2005–2020), Archives, Magdalen College Oxford.
427. Encaenia 2001, 22 June 2001, Supplement to No. 4591, *Oxford University Gazette.*
428. J. Feichtinger, H. Matis, S. Sienell and H. Uhl (Eds), *The Academy of Sciences in Vienna 1938 to 1945* (Austrian Academy of Sciences Press, Vienna, 2014).

429. Le Prix des Trois Physiciens, https://www.phys.ens.fr/fr/article/prix-des-trois-physiciens.

430. American Physical Society, Governance, https://engage.aps.org/dcmp/governance/about.

431. J. D. Martin, When condensed-matter physics became king, *Physics Today*, **72**, 30 (2019).

432. P. W. Anderson, More is different: broken symmetry and the nature of the hierarchical structure of science, *Science*, **177**, 393 (1972).

433. D. Appell, The Supercollider That Never Was, 15 October 2013, *Scientific American*.

434. David Gross, 13 October 1998, Nobel 1998–2001, Box 146, WKP.

435. Arno Penzias — Facts, NobelPrize.org, Nobel Prize Outreach.

436. E. Koch, *Deemed Suspect: A Wartime Blunder* (Methuen, London, 1980).

437. E. Bigley, *Creativity, Community, and Memory Building: Interned Jewish Refugees in Canada During and After World War II*, MA thesis, Carleton University, Ottawa (2017).

438. M. DiChristina, Chemistry Nobelist Attacks Macular Degeneration, 19 July 2012, *Scientific American*.

439. A. Mattsson, 8 December 2016, in KITP Conference: Kohn Science Symposium, Kavli Institute for Theoretical Physics, Santa Barbara, https://online.kitp.ucsb.edu/online/kohnsymposium-c16/.

440. R. Haunschild, A. Barth and B. French, A comprehensive analysis of the history of DFT based on the bibliometric method RPYS, *J. Cheminform.* **11**, 72 (2019).

441. J. P. Perdew, K. Burke and M. Ernzerhof, Generalized gradient approximation made simple, *Phys. Rev. Lett.*, **77**, 3865 (1996).

442. A. D. Becke, Density-functional thermochemistry. III. The role of exact exchange, *J. Chem. Phys.*, **98**, 5648 (1993).

443. P. Verma and D. G. Truhlar, Status and challenges of density functional theory, *Trends Chem.*, **2**, 302 (2020).

444. J. Giles, Software company bans competitive users, *Nature*, **429**, 231 (2004).

445. J. Kong *et al.*, Q-Chem 2.0: a high-performance *ab initio* electronic structure program package, *J. Comput. Chem.*, **21**, 1532 (2000).

446. G. R. Schleder *et al.*, From DFT to machine learning: recent approaches to materials science — a review, *J. Phys. Mater.*, **2**, 032001 (2019).

447. A. Hellmann *et al.*, Predicting catalysis: understanding ammonia synthesis from first-principles calculations, *J. Phys. Chem. B*, **110**, 17719 (2006).

448. J. P. Brodholt and L. Vočadlo, Applications of density functional theory in the geosciences, *MRS Bulletin*, **31**, 675 (2006).

449. D. J. Cole and N. D. M. Hine, Applications of large-scale density functional theory in biology, *J. Phys. Condens. Matt.*, **28**, 393001 (2016).

450. E. N. Muratov *et al.*, A critical overview of computational approaches employed for COVID-19 drug discovery, *Chem. Soc. Rev.*, **50**, 9121 (2021).

451. K. J. Y. Wu *et al.*, An antibiotic preorganised for ribosomal binding overcomes antimicrobial resistance, *Science*, **383**, 721 (2024).

452. R. Pederson, B. Kalita and K. Burke, Machine learning and density functional theory, *Nat. Rev. Phys.*, **4**, 357 (2022).

453. N. Mardirossian and M. Head-Gordon, Thirty years of density functional theory in computational chemistry: an overview and extensive assessment of 200 density functionals, *Mol. Phys.*, **115**, 2315 (2017).

454. M. Karplus, *Spinach On The Ceiling: The Multifaceted Life Of A Theoretical Chemist* (World Scientific Publishing, London, 2020).

455. C. E. Tzeliou, M. A. Mermigki and D. Tzeli, Review on the QM/MM methodologies and their application to metalloproteins, *Molecules*, **27**, 2660 (2022).

456. J. Xiao and B. Yan, First-principles calculations for topological quantum materials, *Nat. Rev. Phys.*, **3**, 283 (2021).

457. S. Roberts, Walter Kohn, Who Won Nobel in Chemistry, Dies at 93, 25 April 2016, *The New York Times*.

458. M. Weil, Walter Kohn, Onetime Refugee Who Became Nobel Laureate in Chemistry, Dies at 93, 24 April 2016, *The Washington Post*.

459. D. Larousserie, Walter Kohn, Prix Nobel de Chimie, Est Mort, 26 April 2016, *Le Monde*.

460. In Memoriam: Dr. Walter Kohn, Kavli Institute for Theoretical Physics, UC Santa Barbara, https://www.kitp.ucsb.edu/news/in-memoriam-dr-walter-kohn.

461. In Memoriam: Walter Kohn 1923–2016, UC San Diego, School of Physical Sciences, https://physicalsciences.ucsd.edu/media-events/articles/2016/0425.html.

462. Remembering U of T's Walter Kohn, University of Toronto, https://www.utoronto.ca/news/remembering-u-t-s-walter-kohn.

463. P. C. Hohenberg and J. S. Langer, Walter Kohn, *Physics Today,* **69**, 64 (2016).

464. Mara Kohn Obituary, 19 December 2018, *Santa Barbara Independent*.

Bibliography

H. G. Adler, *Theresienstadt 1941–45. Das Antlitz einer Zwangsgemeinschaft, Geschichte Soziologie Psychologie (Theresienstadt 1941–45: The Face of a Compulsory Community, History Sociology Psychology)* (Tübingen, Mohr, 1955).

E. J. Allin, *Physics at the University of Toronto 1843–1980* (Department of Physics, University of Toronto, 1981).

H. Arendt, *Eichmann in Jerusalem: A Report on the Banality of Evil* (Viking Press, New York, 1963).

S. Beller, *Vienna and the Jews 1867–1938: A Cultural History* (Cambridge University Press, 1989).

J. Bowle, *Viscount Samuel: A Biography* (Gollancz, London, 1957).

A. Byers, *Saving Children from the Holocaust: The Kindertransport* (Leo Paper Group, Guangdong, 2012).

P. K. Chattaraj (Ed), *Chemical Reactivity Theory: A Density Functional View* (CRC Press, Boca Raton, 2009).

D. C. Clary, *Schrödinger in Oxford* (World Scientific Publishing, Singapore, 2022).

D. C. Clary, *The Lost Scientists of World War II* (World Scientific Publishing, Singapore, 2024).

D. C. Clary and B. J. Orr (Eds), *Optical, Electric and Magnetic Properties of Molecules: A Review of the Work of A. D. Buckingham* (Elsevier, Amsterdam, 1997).

F. Close, *Trinity: The Treachery and Pursuit of the Most Dangerous Spy in History* (Penguin, London, 2019).

J. Craig-Norton, *The Kindertransport: Contesting Memory* (Indiana University Press, 2019).

C. J. Cramer, *Essentials of Computational Chemistry: Theory and Models* (Wiley, Chichester, 2002).

J. A. Cross, *Sir Samuel Hoare: A Political Biography* (Jonathan Cape, London, 1977).

M. Eckert, *Arnold Sommerfeld: Science, Life and Turbulent Times 1868–1951*, T. Artin (Trans.) (Springer, Berlin/Heidelberg, 2013).

J. Eisinger, *Glimpses: A Sundry Life* (Josef Eisinger, 2023).

J. Eisinger, *Flight and Refuge: Reminiscences of a Motley Youth* (Josef Eisinger, New York, 2016).

C. P. Enz, *No Time to be Brief: A Scientific Biography of Wolfgang Pauli* (Oxford University Press, 2002).

G. Farmelo, *The Strangest Man: The Hidden Life of Paul Dirac, Quantum Genius* (Faber and Faber, London, 2009).

J. Feichtinger, H. Matis, S. Sienell and H. Uhl (Eds), *The Academy of Sciences in Vienna 1938 to 1945* (Austrian Academy of Sciences Press, Vienna, 2014).

G. Ferry, *Max Perutz and the Secret of Life* (Chatto and Windus, London, 2007).

T. Frängsmyr (Ed), *Walter Kohn, Les Prix Nobel, The Nobel Prizes 1998* (Nobel Foundation, Stockholm, 1999).

S. Friedländer, *Nazi Germany and the Jews Volume 1. The Years of Persecution, 1933–1939* (HarperCollins, New York, 1997).

O. R. Frisch, *What Little I Remember* (Cambridge University Press, 1979).

K. Gavroglu, *Fritz London: A Scientific Biography* (Cambridge University Press, 1995).

M. Gilbert, *Churchill: A Life* (Henry Holt, New York, 1991).

M. Gilbert, *Kristallnacht: Prelude to Destruction* (Harper, New York, 2007).

P. Gillman and L. Gillman, *Collar the Lot! How Britain Interned and Expelled its Wartime Refugees* (Quartet Books, London, 1980).

M. D. Gordin, *Einstein in Bohemia* (Princeton University Press, 2020).

A. Gottwaldt and D. Schulle, *Die 'Judendeportationen' aus dem Deutschen Reich 1941–1945* (Marix Verlag, Wiesbaden, 2005).

N. T. Greenspan, *Atomic Spy: The Dark Lives of Klaus Fuchs* (Viking, New York, 2020).

N. T. Greenspan, *The End of the Certain World: The Life and Science of Max Born. The Nobel Physicist who Ignited the Quantum Revolution* (Basic Books, New York, 2005).

M. Karplus, *Spinach On The Ceiling: The Multifaceted Life Of A Theoretical Chemist* (World Scientific Publishing, London, 2020).

C. Kerr, *The Gold and the Blue: A Personal Memoir of the University of California, 1949–1997. Academic Triumphs* (University of California Press, 2001).

I. Kershaw, *Hitler: A Biography* (W. W. Norton & Co., New York, 2008).

Y. Kapp and M. Mynatt, *British Policy and the Refugees, 1933–1941* (Routledge, Abingdon, 1997).

M. Levy, *Get the Children Out! Unsung Heroes of the Kindertransport* (Lemon Soul, London, 2023).

M. A. Livingston, *The Fascists and the Jews of Italy: Mussolini's Race Laws, 1938–1943* (Cambridge University Press, 2014).

L. London, *Whitehall and the Jews, 1933–1948: British Immigration Policy, Jewish Refugees and the Holocaust* (Cambridge University Press, 2000).

J. D. Martin, *Solid State Insurrection: How the Science of Substance Made American Physics Matter* (University of Pittsburgh Press, 2018).

K. D. McRae, *Nuclear Dawn: F. E. Simon and the Race for Atomic Weapons in World War II* (Oxford University Press, 2014).

J. Medawar and D. Pyke, *Hitler's Gift: The True Story of the Scientists Expelled by the Nazi Regime* (Arcade, New York, 2000).

J. Mehra and K. A. Milton, *Climbing the Mountain: The Scientific Biography of Julian Schwinger* (Oxford University Press, 2003).

W. J. Moore, *Schrödinger: Life and Thought* (Cambridge University Press, 1989).

P. Neville, *Hitler and Appeasement: The British Attempt to Prevent the Second World War* (Hambledon, London, 2006).

A. Pais, *Subtle is the Lord: The Science and the Life of Albert Einstein* (Oxford University Press, 2005).

J. A. Pople, W. G. Schneider and H. J. Bernstein, *High Resolution Nuclear Magnetic Resonance* (McGraw-Hill, New York, 1959).

D. Rabinovici, *Eichmann's Jews: The Jewish Administration of Holocaust Vienna, 1938–1945*, N. Somers (Trans.) (Polity, Cambridge, 2011).

A. Roberts, *The Storm of War: A New History of the Second World War* (Allen Lane, London, 2009).

L. Rothkirchen, *The Jews of Bohemia and Moravia: Facing the Holocaust* (University of Nebraska Press, 2005).

M. Scheffler and P. Weinberger (Eds), *Walter Kohn: Personal Stories and Anecdotes Told by Friends and Collaborators* (Springer-Verlag, Berlin, 2003).

G. Schneider, *Exile and Destruction. The Fate of Austrian Jews 1938–1945* (Praeger, Westport CT, 1995).

K-H. Schoeps, *Literature and Film in the Third Reich* (Camden House, Rochester, 2004).

F. Seitz, *The Modern Theory of Solids* (McGraw Hill, New York, 1940).

R. Siegmund-Schultze, *Mathematicians Fleeing from Nazi Germany: Individual Fates and Global Impact* (Princeton University Press, 2009).

R. L. Sime, *Lise Meitner: A Life in Physics* (University of California Press, Berkeley, 1996).

A. Speer, *Inside the Third Reich* (Weidenfeld & Nicolson, London, 1970).

J. L. Synge and B. A. Griffith, *Principles of Mechanics* (McGraw Hill, New York, 1942).

A. W. Tarbell, *The Story of Carnegie Tech: Being a History of Carnegie Institute of Technology from 1900 to 1935* (Carnegie Inst. Tech., 1937).

A. Theobald, *Dangerous Enemy Sympathizers: Canadian Internment Camp B, 1940–1945* (Goose Lane, New Brunswick, 2019).

N. Thompson, *The History of the Department of Physics in Bristol: 1948–1988* (Wills Laboratory, Bristol, 1992).

M. Unger (Ed), *The Last Ghetto: Life in the Łódź Ghetto 1940–1944* (Yad Vashem, Jerusalem, 1995).

J. Vinzent, *Identity and Image. Refugee Artists from Nazi Germany in Britain (1933–1945)* (VDG, Weimar, 2005).

H. Von Hofmannsthal, *1553–1953, Vierhundert Jahre, Akademisches Gymnasium, Festschrift* (Akademisches Gymnasium, Wien 1953).

M. Wiesel (Ed), *To Give them Light: The Legacy of Roman Vishniac* (Simon & Schuster, New York, 1993).

V. Weisskopf, *The Joy of Insight: Passions of a Physicist* (Basic Books, New York, 1991).

E. P. Wigner, *The Recollections of Eugene P. Wigner, as told to Andrew Szanton* (Plenum, 1992).

A. Zangwill, *A Mind Over Matter: Philip Anderson and the Physics of the Very Many* (Oxford University Press, 2021).

Permissions

To the knowledge of the author, all photographs are in the public domain unless stated otherwise in the figure captions, where permissions are noted.

The author is very grateful for the following permissions:

Letters, quotes and photographs of Walter Kohn: the family of Walter Kohn.

Comments, quotes, and photographs of Josef Eisinger and friends: Josef Eisinger.

Letters from John Pople: his daughter Hilary Pople.

Letters from Robert Parr: his daughter Jeanne Lemkau.

Photographs of Herbert Neuhaus and friends: his son Peter Neuhaus.

Quotes from interviews conducted by the Vancouver Holocaust Education Centre: Associate Director of Collections and Exhibitions, Caitlin Donaldson.

Email messages from Giovanni Bachelet to Walter Kohn: Giovanni Bachelet.

Photograph of Axel Becke: Axel Becke.

Photograph of Walter Kohn in Athens: Karlheinz Schwarz.

The author also acknowledges helpful assistance from the following individuals, archives and institutions: Department of Special Research Collections, UC Santa Barbara Library, University of California, Santa Barbara (Matt Stahl and Raul Pizano); Special Collections and Archives, UC San Diego (Lynda Corey Claassen); Leo Baeck Institute New York (Elena Butuzova); Yad Vashem Archives (Oxana Korol); Reach Publishing Services (Donella Thompson); Imberhorne School (Angela Nicholls); Science History Institute (Kenton G. Jaehnig); Nate D. Sanders Auctions (Laura Kirk); Bibliothèque et Archives Canada (Cédric Lafontaine); Canadian Mathematical Society (Termeh Kousha); TOBuilt database (Justine Tenzer); Council for At-Risk Academics (Stephen Wordsworth); DÖW/Spanienarchiv

(Manfred Mugrauer); Nobel Prize Outreach (Ulrika Magnusson); Národní archiv (Zdeňka Kokošková); Niels Bohr Archives (Rob Sunderland); Gordon Research Conferences (Mical Mayor); World Jewish Relief (Michelle Werth); Vienna Tourist Guide (Hedwig Abraham); Tyne Built Ships (Dave Waller and Kevin Blair); University of Toronto Chancellor's Office (Brenda Ichikawa); Laura Bialis; John Wonmo Seong; The Royal Society; UK National Archives, Kew; Getty Images; Naczelna Dyrekcja Archiwów Państwowych.

Index

Trois Rivières camp, 42
Tsui, Daniel, 205

U

Ullmann, Viktor, 226, 227
United States Holocaust Museum, 226
University College London, 82
University College Toronto, 59
University of California, 122, 125, 137, 138, 150, 178, 209, 211, 218, 219, 222, 247, 252
University of California, Los Angeles (UCLA), 123, 150, 178
University of California, San Diego (UCSD), 122–125, 130–138, 147–153, 179, 180, 183, 206–209, 222, 224, 225, 252
University of California, Santa Barbara (UCSB), 178–189, 201, 204, 207, 208, 210–213, 236, 248, 252
University of Cambridge, 45, 48, 55, 64, 119, 133, 141, 145, 163, 168–172, 176, 177, 185, 196, 197, 210, 232, 238
University of Chicago, 119, 125, 132, 174
University of Florida, 158, 172
University of Graz, 239
University of Illinois, 95, 97, 112, 148, 173, 205, 206
University of Michigan, 114, 119
University of Oxford, 53, 54, 141, 152, 171, 172, 175, 240, 241
University of Pennsylvania, 118, 119, 124
University of Toronto, 50–69, 92, 105, 108, 143, 151, 180, 202, 203, 209, 241, 244, 252, 253
University of Vienna, 8, 45, 70, 80, 152, 231, 232, 239
Unrestricted Hartree-Fock, 171

Urey, Harold, 123, 129, 130, 132
USSR Academy of Sciences, 224

V

Van Vleck, John, 58, 87–89, 92, 112, 147, 148, 151, 159, 192, 193, 205
Variational Scattering Theory, 90, 103, 113
Veselí, Moravia, 1
Vienna, 1–27, 104, 228
Viennese Handelsakademie, 15
Vishniac, Luta, 182
Vishniac, Mara, 1, 181, 182, 184, 197, 198, 212, 213, 226, 227, 248, 253
Vishniac, Roman, 181, 182, 227, 253
vom Rath, Ernst, 18
von Hindenburg, Paul, 9
von Klitzing, Klaus, 192
von Laue, Max, 192
von Mises, Richard, 8
von Neumann, John, 48
Vosko, Seymour, 121, 133, 156

W

Wallace, Philip, 110
Wannier, Gregor, 112, 115
Warburg, Otto, 51
Warner, John, 119, 125, 126
Warshel, Arieh, 251
Weinberg, Stephen, 223
Weinstein, Alexander, 58, 63–65, 203
Weizmann, Chaim, 21
Welch Award in Chemistry, 194, 195
Welsh, Harry, 58
Werner, Joachim, 196
Weyl, Hermann, 58, 63
Whitehead, J. H. C., 54
Wiesel, Elie, 232
Wigner, Eugene, 66, 93, 96, 97, 132, 138, 174, 192
Wijsmuller-Meijer, Geertruida, 23
Wilczek, Frank, 183

Wilson, Bright, 91, 159, 194
Wilson, Harold, 175
Wilson, Kenneth, 192
Wilson, Robert, 247
Wolf Prize for Chemistry, 137, 177, 195
Woll, Edwin, 133